ECONOMIC ISSUES, PROBLEMS AND PERSPECTIVES

NON-NESTED REGRESSION MODELS

ECONOMIC ISSUES, PROBLEMS AND PERSPECTIVES

Additional books in this series can be found on Nova's website under the Series tab.

Additional e-books in this series can be found on Nova's website under the e-book tab.

ECONOMIC ISSUES, PROBLEMS AND PERSPECTIVES

NON-NESTED REGRESSION MODELS

MAHMOOD A. BODLA

AND

M. ISHAQ BHATTI

New York

Library of Congress Cataloging-in-Publication Data

Non-nested regression models / editors: M. Ishaq Bhatti (IBF Programme, School of Economics and Finance, Faculty of Business, Economics & Law, La Trobe University, Melbourne, Australia) and Mahmood Ahmad Bodla (COMSATS, Institute of Information Technology, Lahore, Pakistan).
pages cm
Includes bibliographical references and index.
ISBN 978-1-62417-770-5 (soft cover)
1. Econometric models. 2. Regression analysis. I. Bhatti, M. Ishaq. II. Bodla, Mahmood Ahmad.
HB141.N653 2013
519.5'36--dc23

2012050345

Published by Nova Science Publishers, Inc. † New York

CONTENTS

PREFACE

The book consists of recent development in non-nested regression and reviews some of the current literature and contributes in non-nested tests based on regression models. It addresses two interrelated problems in economics modeling: non-nested hypothesis testing in econometrics and regression models with stochastic regressors. The primary motivation for this book stems from the nature of econometric models. As an abstraction from reality, each statistical model consists of mathematical relationships and stochastic, behavioral assumptions. In practice, the validity of these assumptions and the adequacy of the mathematical specifications are ascertained through a series of diagnostic and specification tests. Conventional test procedures, however, fail to recognize that economic theory generally provides more than one distinct model to explain any given economic phenomenon. This book also contains some of the published work of the authors - Bodla and Bhatti (2007, 2013) and Bhatti and Bodla (2008).

ACKNOWLEDGMENT

We acknowledge the patience of our families while we were busy in finalizing this book particularly, Talat Afza Bodla and Asma Bhatti for their cooperation and Daniel Nguyen for excellent research Assistantship.

Chapter 1

INTRODUCTION

1.1. PREAMBLE

This book intertwines two important issues of applied econometrics: non-nested hypothesis testing and regression models with stochastic regressors. The primary motivation for this book stems from the very nature of econometric models; as an abstraction from reality, each statistical model consists of mathematical relationships and stochastic, behavioral assumptions. In practice, the validity of these assumptions and adequacy of the mathematical specifications is ascertained through a series of diagnostic and specification tests. Conventional test procedures, however, fail to recognize that economic theory generally provides more than one distinct model to explain any given economic phenomenon.

Several empirical studies, for example, involve modeling of money demand as assets or transactions models; unemployment theory leads to Keynesian versus neo-classical models; specification of production technologies includes constant or variable elasticity of substitution vis-à-vis translog functions; time series models may involve autoregressive or moving average processes. Such competing models are separate or non-nested in the sense that neither model can be obtained as a limiting or special case of the other. In general, conventional tests either inadequately test or lack power to test non-nested hypotheses. Thus, following Cox (1961, 1962) and Davidson and MacKinnon (1981), various extensions of likelihood ratio and artificial embedding procedures have been developed to test non-nested, single and multivariate regression models involving distinct regressors, functional forms, or error structures.

These specialized non-nested tests typically apply to regression models individually satisfying the full set of classical regression model assumptions. But in applied econometrics, it is common to expect at least one assumption violation - regression analysis of time series generally includes identification of the order of autoregressive error structure; modeling of systems of demand equations may involve "seemingly unrelated" correlations of the disturbances. In addition, the assumption of non-stochastic regressors seldom holds in social sciences, where researchers, typically have no control over the experimental design, construction, or collection of survey information. Such situations generally lead to economic variables that are unobserved or observed with error. To this extent, the "truth" of one of the two non-nested regression models cannot be seriously maintained in applications to economic phenomena.

The violation of classical assumptions, therefore, requires special consideration in statistical inference. Indeed, least squares estimation is replaced with full information maximum likelihood and instrumental variable estimation procedures to deal with the violation assumptions of various types. Of equal importance is the study of potential distortions to the sampling distributions of various test statistics. Recently, Bhatti (2012) considers testing for cluster effects using tested framework. This book will concentrate only on non-nested regression based testing issues.

The purpose of this book is to examine the robustness and sensitivity of non-nested hypothesis tests in single equation, mismeasured regression models. The study of asymptotic properties and Monte Carlo evidence of small sample behavior of these tests is expected to provide a significant contribution to the econometrics literature for two reasons. Firstly, the presence of non-nested regression models is not limited to the examples cited above; rather, there may be as many empirical applications of non-nested models in economics as nested ones. Therefore, the study of these relatively new statistics under various scenarios is likely to be of widespread interest to many empiricists. Secondly, the presence of stochastic regressors due to either measurement errors or unobservable nature of economic variables may well be the "rule rather than an exception". Some typical examples of mismeasured or unobserved explanatory variables in several sub-disciplines of economics are: the concept of utility in microeconomics, permanent income in macro-economics, various measures of interest rates in money and banking, market concentration measures in industrial organization, qualitative measures of location variables in urban economics, and human capital concepts such as ability, education, experience, and achievement in labor economics. In this sense, the two major econometric issues addressed here, non-nested hypotheses and mismeasured variables may occur together with some frequency in applied economics.

The remainder of this chapter provides preliminary details such as the specification and estimation of errors-in-variables models (EVM) and hypothesis testing procedures of various types of non-nested test statistics used in the proceeding chapters.

1.2. MODEL SPECIFICATION

Economic models emerge through a process of abstraction from complex real world economic phenomena. The true relationship among various economic variables is seldom known with certainty. Thus, several possible explanations, or more formally economic models, may be postulated to approximate these relationships. According to McAleer and Deistler (1986):

> By model specification is meant the set of all assumptions, which define the model class, and hence also the parameter space for the inference procedure. (McAleer and Deistler 1986, pg. 5)

As an example, consider the standard textbook model specification which relies on the following classical assumptions.

A1) $y = X\beta + u$.
A2) X is a non-stochastic and finite matrix of rank k.
A3) $X'X$ is non-singular for all n.
A4) E(u) = 0, where E denotes the expectation operator.
A5) u is independently and identically distributed as a normal

Note that in the above set of assumptions (from A1 to A5), the disturbance variate vector, u in assumption A1 is an unobserved random error which assumed to be normally distributed with zero mean and homocedastic variance i.e., $u \sim N(0, \sigma^2 I_n)$, where σ^2 is a finite, unknown scalar. Moreover, y is an nx1 vector of observations on the dependent variable and n denotes the number of observations in the sample. The rest of the components of the model are; X with nxk matrix of explanatory, non-stochastic variables of full column rank and β is kx1 vector of unknown position parameters.

This "classical normal linear multiple regression model" provides the simplest specification to estimate the parameters consistently by using Ordinary least squares (OLS) estimation and conduct classical hypothesis tests. However, the validity of this model cannot be guaranteed if the underlying assumptions are not consistent with the real world economic phenomenon. Contrary to the above assumptions, most practical situations involve non-linearity in the parameters, parameters varying with observations, autoregressive or heteroscedastic error terms, and so forth. There have been a number of advances in the econometric literature to diagnose such violations of classical assumptions and design alternative estimation strategies.

The present study focuses on the problem of errors in economic variables which violates the non-stochastic nature of X in assumption (A.2) above. For this reason, the following section discusses the problems of stochastic regressors, errors-in-variables and specifications involving mismeasured or unobserved variables.

1.3. Fixed Verse Stochastic Regressors

The assumption of non-stochastic regressors may well be valid for the experimental data produced, for example, in the physical sciences, where observations can be fixed from one sample to the other within a controlled environment. On the other hand, economic data are not usually produced by the researcher according to some experimental design. Rather, they are the product of economic agents who may be acting systematically or randomly according to their individual behavioral characteristics. Thus, as a data generating process, the random behavior of economic agents provides for the possibility of stochastic variables. On the other hand, various practical problems related to data measurement or our inability to observe some economic variables are also major causes of stochastic variables. For example to measure farm efficiency, the agriculture output depends on the stochastic coefficients associated with random effect due to cultivated land size, various seed quality and the method of cultivation which vary from one area to other and hence constant regression coefficient may not be suitable to measure the efficiency, therefore the use of stochastic model is preferred over fix (See Chapter 7 Bhatti (2012)).

The existence of stochastic regressors in linear regression models poses a serious problem for estimation since the OLS estimators are biased and inconsistent. However, the presence of stochastic regressors in the linear regression model above will not cause any econometric problems if an additional assumption regarding the statistical independence of the error terms and regressors can be made. For this reason, the following assumption is sometimes formally included in the above regression model.

A6) $E(u|X) = 0$ where 0 is an nxl vector of zeros.

An alternative solution to the problem of inconsistency in such situations is to obtain instrumental variable (IV) estimators for β. In this case, a matrix, W, of order nxk is defined as the IV matrix for X if it satisfies the following assumptions in large samples.

(i) $Cov(u, W) = 0$
(ii) $Cov(X, W) \neq 0$
(iii) W'W is non-singular.

Then the IV estimator for β may be defined as

$$\beta = (X' A_w X)^{-1} X' A_w y, \text{ where } A_w = W(W'W)^{-1} W'$$
$$= \beta + (X' A_w X)^{-1} X' A_w u$$

The consistency of $\widetilde{\beta}$ simply follows from the asymptotic independence of W and u, a formal proof of which is given in chapter three.

In practice, the violation of $E(u|X) = 0$ is very common in economics. Illustrations of unobserved or mismeasured variables are now discussed in some detail.

1.4. Errors-in-Variables Models

Consider a simplified econometric specification of the consumption function,

$$C = \alpha + \beta I + u,$$

where C and I represent observed values for consumption and income, respectively, u is a vector of unobserved random errors, and a and β are unknown parameters to be estimated. In the classical regression framework, I is assumed to be either non-stochastic or uncorrected with the error terms, but, according to Friedman (1957), $C = C_p + C_t$ and $I = I_p + I_t$ where subscripts p and t refer to unobserved permanent and transitory components respectively.

The permanent components of income and consumption generally vary systematically with individual behavioral characteristics but their transitory counterparts change randomly. Thus, the observed quantities cannot be correctly measured, in essence, due to the unobservable components.

Observed values may also be subject to errors from various data collection and surveying problems. Griliches (1974) characterizes the problem of errors-in-variables are as follows:

> Much of the problem, I think, arises because of the separation in economics between data producers and data analyzers ... Our data are largely reported by firms and persons who are not professional observers and who do not have any stake in the correctness and precision of the observations they report. While economists have increased their use of surveys in recent years and commissioned a few special purpose ones of their own, in general, the data collection and thus the responsibility for the quality of the collected material is still largely delegated to census bureaus, survey research centers, and similar institutions, and is divorced from the direct supervision and responsibility of the analyzing team. It is also fair to note that part of the problem arises from our own fuzziness about what it is we would like to observe and from the complexity of the phenomena which we are trying to measure. (Griliches 1974, pg. 973-974)

Errors in variables are not caused solely by measurement problems in the observed magnitudes. Rather, there are several economic variables such as utility, ability, achievement, permanent income, expected prices, and reservation wages that are, by their very nature, not directly observable. Thus, the term errors-in-variables in this study will refer to mismeasured as well as unobserved variables. It is well known that inconsistency of the least square estimators arises from mismeasurement of the right hand side variables only. For the reason, this book will not emphasize the problem of mismeasured regression. However, consistent estimation of econometric models is not an end itself; rather, rigorous testing of underlying assumptions is needed to enhance the validity of predictions from such models. Thus, the next section provides a brief discussion of various econometric testing procedures widely used in economics.

1.5. Model Testing and Model Selection

The specification of econometric models involves a number of testable assumptions; the true data generating process (DGP) represented by any econometric model is too complex to be specified exactly. Thus, several hypothesis tests are performed using observed or simulated sample observations. It should be noted that the "true values of parameters" or "true models" or "true DGP" are purely theoretical concepts that, nonetheless, play a very significant role in developing the test criteria. Just recently, theoretician and practitioners are engaged in discussing controversial issue whether to use hypothesis testing approach or model selection approach in deciding the best models which fit well to DGP approach. Bhatti, et al. (2006) reviewed both approaches and observed that the DGP approach fits well.

There are two basic approaches to statistical inference: classical and Bayesian. Classical methods are widely practiced and may be further classified into three categories: diagnostic or misspecification tests, specification tests, and non-nested tests. Each of these tests and the Bayesian approach are discussed in the following subsections.

1.5.1. Diagnostic or Misspecification Tests

Any econometric model, for example the one specified in section 1.1, represents a set of assumptions and, until tested against the sample evidence, the validity of such assumptions remains questionable. One measure of the discrepancy between the true DGP and the estimated models is the size of the residual terms.

In fact, residuals are frequently used to diagnose various econometric misspecifications such as parametric non-linearities, heteroscedasticity, autoregression, structural instability, serial correlation, equi-correlated or familial correlated residuals and cluster effects in complex surveys (Bhatti 2012).

To conduct diagnostic tests, some well specified "null" hypothesis are compared with another general "alternative" hypothesis to see the validity of the model. For example, the null hypothesis of constant variance may be tested against the alternative of heteroscedasticity without specifying the exact form or source of heteroscedasticity.

Similarly, the hypothesis of no serial correlation can be evaluated against the presence of serial correlation without any particular designation of the autoregressive (AR) or moving average (MA) processes. Thus, based on some measure of "goodness of fit", any significant departure away from the null hypothesis can be observed, implying that the null hypothesis will be rejected as the true specification.

However, significant departures may be sometimes induced by problems other than those specified in an alternative model. For example, a significant departure from the null hypothesis of constant variance may be caused by an incorrect functional form rather than heteroscedasticity. Thus, to diagnose the correct form of misspecification, it is always desirable to conduct a series of specifications tests. Even then the hypothesis testing approach may not be appropriate in some application and one may consider using model selection procedures to decide the best model for specific data structure.

1.5.2. Specification Tests

Unlike the diagnostic or misspecification tests, specification tests are employed for specific alternative hypotheses. However, in the nested case, the null hypothesis (H_0) is still obtained as a restricted form of alternative hypothesis (H_1). In the classical framework, the entire sample space is partitioned into two mutually exclusive regions; one is the acceptance region, where the DGP is consistent with H_0, and the other will be the rejection or "critical" region. An arbitrary choice of the critical region increases either the probability of rejecting a true H_0 (*Type-I error*) or the probability of accepting a false H_0 (*Type-II error*). However, for some "optimal" critical region, the probability of a Type-II error can be evaluated in the presence of a specific alternative hypothesis. Thus, in the case of specification tests it may be possible to obtain a uniformly most powerful test; *power* here refers to the probability of not committing Type-II error. Specification tests can be classified into Likelihood Ratio (LR) tests, Lagrange Multiplier (LM) tests, or Wald (W) tests. The ratio of maximized likelihood functions under the null and alternative models is the basis of the LR statistic which follows a well-defined distribution under the null hypothesis when H_0 is nested within H_1. Once the sampling distribution of a test has been determined, it is generally straightforward to set up the critical region and evaluate its power.

Specification tests based on LR principle require estimation of both null and alternative models. Alternatively, tests based on the LM and W procedures require estimation of only one of the two models; W tests are obtained by evaluating the unrestricted model, H_1, while the LM procedure requires estimation of H_0 only. In practice, the choice of testing procedures will generally depend on computational convenience. These tests are asymptotically equivalent to each other under H_0, although they will differ, of course, in finite samples. The Monte Carlo evidence of Berndt and Savin (1977) and Breusch (1979) suggests the following inequality under special circumstances:

$$W \geq LR \geq LM.$$

It is important to note that the above three tests have been proved to be asymptotically equivalent under specific situation.

1.5.3. Non-Nested Tests

Classical statistical inference arises when the null (or restricted) model is nested in a more general (or unrestricted) alternative model. In practice, however, an cconomic theory may be involve in competing econometric specifications which are in non-nested format, in the sense that neither model can be expressed as a special case of the other. More formal definitions and some illustrative examples of non-nested models may be found in the next section.

Traditional F-tests may be applied to linear non-nested models when the null and alternative models are embedded in an artificial compound model. Each model can then be tested as a set of restrictions on this compound model. However, this procedure generally results in poor power performance of the tests under consideration.

Cox (1961, 1962) provided the pioneering work in non-nested testing by modifying the LR principle. Several other tests, which are computationally simpler than, but asymptotically equivalent to the Cox-test, have more recently been devised. Details of these and other procedures may be found in the next two chapters.

1.5.4. Bayesian Approach to testing

From a Bayesian viewpoint, testing nested or non-nested models is based on a comparison of posterior odds of the competing models. The computation of posterior odds involves the following.

1) The null (H_0) and alternative (H_1) are assigned certain probabilities, termed *prior probabilities*, based on the researcher's subjective knowledge about the DGP prior to any sample information.
2) Prior probabilities and the sample information are then combined together into a likelihood function with the help of Bayes' theorem to yield the *posterior probabilities*.

If the posterior probability of H_0 exceeds the posterior probability of H_1, then H_0 is accepted as the best model. However, except for the case of non-nested models, it is not always possible to accept or reject the competing hypotheses. Thus, the basic purpose of the Bayesian approach according to Zellner (1971) is to 'compare' rather than 'test' the hypotheses of interest.

In contrast to the classical approach, the Bayesian approach has been strongly commended by Zellner (1984, pg. 187-200), but Pesaran (1974) argues that it lacks appeal due to the following fundamental shortcomings.

(i) The outcome of any test can be strongly influenced by the subjective judgment of prior probabilities.
(ii) The restrictive requirement of Bayes' theorem that the models under consideration should be mutually exhaustive is seldom true in practice.

For these reasons, the present study pursues the classical framework to conduct non-nested tests.

1.6. Nested Versus Non-Nested Models

It has been noted earlier that several models may be postulated to approximate the true DGP. In general, the alternative specifications may not belong to the same parametric family; in these instances, the models are said to be separate or non-nested. Two models were defined earlier to be non-nested when neither can be obtained as a limit or special case of the other; conversely, one model is nested within another. To formalize this definition, consider an independently identically distributed (iid) normal variable y such that:

H_0: Probability density $f(y, \beta)$
H_1: Probability density $g(y, \gamma)$

where β and γ , unknown parameter vectors, are contained in the respective parameter spaces B and Γ .

Quandt (1974, pg. 92) defines the hypotheses H_0 and H_1 to be non-nested if $B \cap \Gamma \neq B$ and $B \cap \Gamma \neq \Gamma$. However, to obtain more formal definitions of nested and non-nested models, Pesaran (1987, pg. 70) notes that "...a suitable measurement of closeness of the hypotheses in question is needed".

The Kullback and Leibler (1951) information criterion (KLIC) for discrimination between H_0 and H_1 is obtained as the expected value of the log likelihood ratio of two densities under H_0, i.e.,

$$I_{01}(\beta,\gamma) = \int R_0 \log(f(y,\beta)/g(y,\gamma))\, f(y,\beta)\, dy$$

where R_0 represents the range of y under H_0. The "closeness" of H_1 to H_0 may be then defined as a minimum value of I_{01} over γ, $C_{01}(\beta_0) = \min_\gamma I_{01}(\beta_0,\gamma)$. Here, β_0 is the true value of β under H_0. The closeness of H_0 to H_1, denoted as C_{10}, may be defined analogously. Thus, in terms of C_{01} and C_{10}, Pesaran (1987, pg. 72) provides the following definitions of nested and non-nested hypotheses:

DEFINITION 1 (Nested hypotheses)

The hypothesis H_0 is nested within H_1 if and only if $C_{01}(\beta_0) = 0$ for all admissible values of β_0, while H_1 is nested within H_0 if and only if $C_{10}(\gamma_0) = 0$ for all admissible values of γ_0.

DEFINITION 2 (Globally non-nested hypotheses)

The hypotheses H_0 and H_1 are globally non-nested if $C_{01}(\beta_0)$, and $C_{10}(\gamma_0)$ are both nonzero for all admissible values of β_0 and γ_0.

DEFINITION 3 (Partially non-nested hypotheses)

The hypothesis H_0 is partially non-nested with respect to H_1 if $C_{01}(\beta_0) \neq 0$ for some but not all the admissible values of β_0 . Similarly, H_1 is partially non-nested with respect to H_0 if $C_{10}(\gamma_0) \neq 0$ for some but not all the admissible values of γ_0.

To illustrate these definitions, consider the empirical problem of selecting appropriate scale and opportunity cost variables in money demand models. The following specifications of money demand (MD_t) involve current income (YC_t), permanent income (YP_t), and wealth (W_t) as scale variables and corporate bond rate (BR_t) and commercial paper rate (PR_t) as measures of opportunity cost. Based on the above independent variables, the following hypothesis can be constructed;

$$H_0: MD_t = \alpha_0 + \alpha_1 YC_t + \alpha_2 PR_t + u_t$$
$$H_1: MD_t = \beta_0 + \beta_1 YP_t + \beta_2 BR_t + v_t$$
$$H_2: MD_t = \gamma_0 + \gamma_1 W_t + \gamma_2 BR_t + w_t$$
$$H_3: MD_t = \tau_0 + \tau_1 YC_t + \tau_2 YP_t + \tau_3 BR_t + \tau_4 PR_t + e_t$$

It's important to note that the above hypothesis, H_3 nests to both H_0 and H_1; for example, H_3 collapses or to become a special case to H_0 when under the null hypothesis the parameters tau take value zero i.e., under H_0: $\tau_2 = \tau_3 = 0$.

On the other hand, models H_0 and H_1 are globally non-nested since both include complete sets of distinct regressors, as are H_0 and H_2. However, H_1 and H_2 (involving the choice of appropriate scale variable) or H_2 and H_3 represent partially non-nested hypotheses, since both pairs have at least one common and one non-overlapping variable.

To further illustrate the concept of non-nested models, some typical examples from the econometric literature are discussed below.

Example 1: *Regression Models with Non-Overlapping Explanatory Variables*

The most frequently cited example of non-nested models consists of the following linear regression models under standard assumptions detailed in section 1.2.

$$H_0: y = X\beta + u_0, \qquad u_0 \sim N(0, \sigma_0^2 I_n)$$

$$H_1: y = Z\gamma + u_1, \qquad u_1 \sim N(0, \sigma_1^2 I_n)$$

where in the above null (H_0) and alternative (H_1) hypothesis, X and Z are matrices with at least one distinct column each. Some empirical examples from the existing economics and finance literature are from Johannes and Nasseh (1985), Thornton (1985), investment behavior specifications by Wisely and Johnson (1985), and unemployment models by Pesaran (1982a) are among others.

Example 2: *Regression Models with Different Functional Forms*

Regression models will be non-nested if the functional forms cannot be transformed into one another through parametric restrictions. Gregory and

McAleer (1983) consider the following non-nested functional forms of money demand models.

H_0: $y = \beta_0 + \sum_i \beta_i X_i + u_0$ "Linear"

H_1: $\ln y = \gamma_0 + \sum_i \gamma_i \ln X_i + u_1$ "Log-Linear"

H_2: $y = e^{y_0} \prod_i X_i \gamma_i + u_2$ "Cobb-Douglas"

Example 3: *Regression Models with Different Error Structures*

The existence of competing non-nested models may be caused by alternative specifications of the error terms such as

H_0: $y = e^{\beta_0} X^{\beta_1} + u_0$ "Additive error"

H_1: $y = e^{\gamma_0} X^{\gamma_1} e^{u_1}$ "Multiplicative error"

See, for example, Aneuryn-Evans and Deaton (1980) and non-nested time series models by King and McAleer (1984):

H_0: $y_t = \sum \beta_i y_{t-i} + e_t$ "AR(p)"
H_1: $y_t = \sum \gamma_i e_{t-i} + v_t$ "MA(q)"

where AR(p) is the pth order of AR process and MA(q) is the qth order of MA process in time series data structure.

Example 4: *Probability Distributions*

Much of the major works on non-nested tests deal with univariate probability distributions belonging to separate parametric families. Examples of such distributions are given below:

1. Lognormal versus exponential or gamma or Weibull or Dirichlet distributions. In the literature various studies are done which involve the comparison of some of these distributions. For example the work of Cox (1961, 1962), Jackson (1968), Atkinson (1970), Pereira (1977, 1978), Epps, Singleton, and Pulley (1982), Shen (1982), and Pesaran (1984) are among others.
2. Weibull versus gamma is considered by Pereira (1977).

3. Poisson versus geometric is studied by Cox (1962) and Atkinson (1970).
4. Normal versus logistic is tested by Morimune (1979).

1.7. Outline of the Book

The major focus of this book has been described at the onset as the analysis of non-nested testing procedures when the null, alternative, or both models are specified with stochastic regressors. The above examples illustrate that many economic propositions can be modeled and tested within a non-nested framework. Such non-nested models may consist of mismeasured or unobserved regressors. In such instances, the behavior of non-nested tests in terms of asymptotic power, as well as small sample behavior, needs to be evaluated. To attain these goals, the remainder of the book is organized in the following fashion.

The second chapter summarizes the recent developments in the non-nested regression literature including non-nested models and the tests. It discusses the analytical evaluation of non-nested tests, asymptotic techniques and methodologies used. Chapter three provides an introduction to asymptotic theory to non-nested tests and extends the idea of chapter two. Chapter four presents the fundamental analysis of non-nested tests for several types of errors-in-variables specifications. The existing non-nested tests are modified in such a way that they provide theoretical basis for the construction of some consistent tests for the EVM models. The asymptotic efficiency and power function of various tests are compared under the fixed and local alternative approaches. Chapter five provides finite sample evidence on the performance of various non-nested EVM tests using Monte Carlo method. Chapter seven demonstrates an empirical power comparison of the selected non-nested tests to show relative strength of various tests. The final chapter concludes the book with a summary of major findings.

Chapter 2

RECENT DEVELOPMENT IN NON-NESTED REGRESSION

2.1. INTRODUCTION

Typical hypothesis tests are conducted within a nested framework, where the null hypothesis is obtained as the limit of more general alternative hypotheses. In practice, however, it is plausible that economic theory may generate separate (or non-nested) null and alternative hypotheses. When nested tests statistics are used for the testing of non-nested models, the resultant conclusions are inappropriate. The statistical inference and power of the tests are under estimated with lower power. A formal testing approach to non-nested hypotheses, within the classical framework, emerges with the seminal work of Cox (1961).

Cox (1961) demonstrated a number of criterions to test separate hypotheses such as the Bayesian approach (refer subsection 1.4.4 in chapter 1), Barnard's (1959) procedure involving the mean likelihood ratios, the artificial embedding approach of Hoel (1947), and the modified likelihood ratio approach. The major focus of Cox (1961) is the modified likelihood ratio approach, while Atkinson (1970) uses the principle of exponential embedding, and Quandt (1974) concentrates on the linear embedding of separate null and alternative hypotheses. Following these earlier developments, a large number of specialized tests have been developed to test a null hypothesis against single or multiple non-nested alternative hypotheses. The discussion of these procedures and various theoretical and simulated applications has been organized into the following sections.

The next section describes the classical methods of statistical inference (i.e., testing and estimation) and it elaborates on the limitations that arise when these procedures are applied to non-nested models. The modified likelihood ratio approach of Cox (1961) and its extensions to regression models are discussed in the next section. The third and fourth sections present various non-nested tests based on artificial regressions and instrumental variables respectively.

According to Lodewyckx et al. (2011) non-nested models are not formed from incremental developments of the same theory, but originated from very different hypotheses. The widely used Bayes factor, as intuitive and principled model selection tool from Bayesian statistics, compares competing models such as non-nested and hierarchical models.

There are two main approaches to construct non-nested tests: encompassing approach (Mizon, 1984; Mizon and Richard, 1986) and the comprehensive-model approach (Atkinson, 1970). As regards former approach, Kuan and Lin (2010) pointed out that the existing encompassing tests could not be readily applied to the models, such as quantile regression (QR) of Koenker and Bassett (1978). Therefore, Conditional Quantile Encompassing (CQE) test for non-nested and linear QR models were proposed. Results of simulation confirmed that the proposed tests had an asymptotic $\chi 2$ distribution and performed well in finite samples. In addition, CQE test was found to be asymptotically equivalent to regression rank score test in a comprehensive model under conditional homogeneity which was consistent with the argument of Bontemps and Mizon (2008). It is asserted that there is implicitly a comprehensive model underlying any tests obtained from the encompassing approach.

Kelejian (2008) suggested a J-test for testing a null model containing spatial lags in both dependent variables and disturbance terms against a set of alternative non-nested models that may, but not necessarily, relate solely to the specification of the weighting matrix. Although the proposed test was computationally simple and intuitive, not all of the available information in efficient fashion was employed to determine predictions. Likewise, Kelejian and Piras (2011) presented modification in the J-test to address the shortcomings of Kelejian (2008)'s suggested test. Monte Carlo model was specified in which experimental design based on the format extensively used in studies on spatial regression models (e.g. Arraiz et al., 2010; Piras and Lozano-Gracia, 2010) and in studies on non-nested tests (Pesaran, 1982; Davidson and MacKinnon, 1981; Godfrey and Pesaran, 1983; Delgado and Stengos, 1994). The results indicated that a proper use of the information increased the power of the tests relative to the original J-tests suggested in Kelejian (2008). Contrary to the intuition, results from large and small Monte Carlo samples indicated that J-test procedure could not be employed to establish a test for the structure of the error term.

Although empirical applications of Bayesian methods are abundant (e.g. Clark and McCracken, 2009; Wright, 2008), little is understood about their theoretical sampling properties providing a starting point to explore asymptotic relation between Bayesian model averaging and frequentist predictor selection in both nested and non-nested models (Hong and Preston, 2011). In this regard, Hong and Preston (2011) derived conditions under which their difference was of a smaller order of magnitude than the inverse of the square root of the sample size in large samples. Weak conditions were given under which consistent model selection was feasible, without considering whether the model was nested or non-nested and correctly or incorrectly specified. Bayesian posterior odds and BICs were found consistent for selecting among nested models, but inconsistent for selecting among non-nested models.

Fernandez et al. (2010) analyzed how well several alternative Taylor-rule specifications describe Federal Reserve policy decisions in real time, using the new test developed Giacomini and Rossi (2007) for non-nested model selection in the presence of possible parameter instability. Examination of several alternative Taylor-rule specifications strongly suggests that a version of the Taylor rule needs to be adopted to include both gradualism and preemption in order to describe Federal Reserve funds-rate decisions well consistently.

Korenok (2008) compared empirically two non-nested models such as sticky information and sticky price models for relationship between the world war II, the US aggregate price and labor share by using Bayesian full information likelihood approach. Estimation methodology differed from two-step approach used by Mankiw and Reis (2001), Sbordone (2002), Khan and Zhu (2002), and Kiley (2007). In the first step, a driving process was estimated and then incorporated into a linear rational expectations solution for process of interest. The sticky price model dominated sticky information model, however, an unrestricted VAR dominated both models.

Otsu et al. (2011) proposed non-nested hypothesis tests, Kolmogorov–Smirnov and Cramér–von Mises type moment encompassing tests, for conditional moment restriction models based on the method of generalized empirical likelihood (GEL). They derived the null distributions, validity of a bootstrap procedure, in addition to local and global power properties of designed tests. Simulation results suggested that the tests had reasonable size and power performance in finite samples.

The so-called encompassing principle, which unifies the diversified literature on non-nested tests, is briefly outlined in the fifth section. Empirical applications and statistical properties of non-nested tests are summarized in section six and seven respectively. Conclusions and a brief summary may be found in section eight.

2.2. THE CLASSICAL TESTS AND NON-NESTED HYPOTHESES

Classical procedures based on the likelihood ratio, Wald, and Lagrange Multiplier (or Score) principles are available for the testing of nested hypotheses. As an example, consider the following possibly non-linear regression models:

$$H_0: y = f(X, \beta) + u_0, \quad u_0 \sim N(0, \sigma_0^2 I_n) \tag{2.1}$$

$$H_1: y = g(Z, \gamma) + u_1, \quad u_1 \sim N(0, \sigma_1^2 I_n) \tag{2.2}$$

where in (2.1) the parameters $(\beta, \sigma_0^2) \in \Omega_0$ and in equation (2.2), $(\gamma, \sigma_1^2) \in \Omega_1$ which are unknown shape and position parameters. Given n sample observations for y, X and Z, the likelihood ratio statistic is written as:

$$-2\,[L_0(\hat{\Omega}) - L_1(\hat{\Omega}_1)] \tag{2.3}$$

where $L_i(\hat{\Omega}_i)$ denotes the maximized log likelihood value for model i = 0, 1, under null and alternative hypothesis (2.1) and (2.2), respectively.

If H_0 is nested within H_1, then this likelihood ratio follows a well-defined χ^2 distribution under H_0 and also provides the uniformly most powerful (UMP) critical region for a simple H_0 against a simple H_1. In certain situation the resultant test is a non-central F- Distribution, (see Bhatti 2012, chapter 3, for an example of UMP test). However, when H_0 and H_1 are non-nested, the likelihood ratio does not follow any well-known distribution.

Alternatively, when H_0 and H_1 are linear in β and γ, the classical F-test may be used. This test is based on the artificial compound model:

$$H_F: y = X\beta + \bar{Z}\gamma_1 + u, \tag{2.4}$$

where $\bar{Z}$ in (2.4) consists of r column vectors which are not linearly related to any columns within the matrix X, and γ_1 are the associated parameters in γ. The F-test, for $\gamma_1 = 0$, which is known to be asymptotically equivalent to the Lagrange Multiplier (LM) test, may be written as:

$$F_0 = \frac{y'M_X\bar{Z}(\bar{Z}'M_X\bar{Z})^{-1}\bar{Z}'M_Xy}{\hat{\sigma}_0^2} \tag{2.5}$$

where $M_x = I - X(X'X)^{-1}X'$, and $\hat{\sigma}_0^2$ is a consistent estimator for σ_0^2. Under H_0, statistic F_0 is asymptotically distributed as a central χ^2 with r degrees of freedom. To test H_1 against H_0, a similar statistic may be obtained by interchanging the roles of H_0 and H_1. Although this F-test provides the simplest way to test non-nested models, its declining degrees of freedom with increasing number of non-overlapping variables in $\bar{Z}$ can reduce its power considerably (Pesaran, 1982). For these reasons, it may be desirable to consider the use of separate hypothesis tests when H_0 and H_1 are non-nested.

2.3. The Modified Likelihood Ratio

In his pioneering work on separate hypothesis testing, Cox (1960, 1962) modifies the likelihood ratio statistic of equation (2.3) so that a test based on some known sampling distribution is obtained for non-nested hypotheses. The Cox test and its applications to econometric regression models are discussed in the following two subsections.

2.3.1. The Cox Test

Cox (1961) modifies the likelihood ratio test by evaluating $L_0(\hat{\Omega}_0) - L_1(\hat{\Omega}_1)$ around its expected value under H_0 and obtains the following test statistic for H_0 against H_1:

$$T_0 = [L_0(\hat{\Omega}_0) - L_1(\hat{\Omega}_1)] - E_{\hat{\Omega}_0}[L_0(\hat{\Omega}_0) - L_1(\hat{\Omega}_1)], \tag{2.6}$$

where in (2.6) the expectation under H_0, is evaluated at given value of $\hat{\Omega}_0$. In practice, however, the expectation operation is replaced with probability limit.

Subsequently, Cox (1962, pg. 408-411) shows that the statistic T_0, under H_0, is asymptotically normal with zero mean and finite variance V_0. White (1982) provides a further discussion of the regularity conditions to formalize the distributional structure of the Cox test. However, ascertaining the significance of H_1, on the basis of T_0 alone is not possible; rather, a similar test statistic T_1 is constructed, which is given below:

$$T_1 = [\, L_1(\hat{\Omega}_1) - L_0(\hat{\Omega}_0)\,] - E_{\hat{\Omega}_1}[\, L_1(\hat{\Omega}_1) - L_0(\hat{\Omega}_0)\,] \qquad (2.7)$$

Earlier studies such as Dhrymes et al. (1972, pg. 297), Gaver and Geisel (1974, pg. 59), Pesaran (1974, pg. 158), Ramsey (1974, pg. 37), Harvey (1977, pg. 465), Aneuryn-Evans and Deaton (1980, pg. 289-290), Harvey (1981, pg. 179), Green, Pope and Phipps (1981, pg. 402), and Gourieroux, Monofort and Trongon (1984) tried to draw inferences about the validity of H_1 from T_0 alone. But Fisher and McAleer (1979) and Dastoor (1981) provided a correct interpretation of the Cox test and noted that, for a given set of observations, T_0 and T_1 can jointly lead to only one of the following possible outcomes.

a) Reject both H_0 and H_1 as true models if T_0 and T_1 are significantly different from zero.
b) Accept both H_0 and H_1 as true models if T_0 and T_1 are both insignificant.
c) Accept H_0 and reject H_1 as true models if T_0 is insignificant and T_1 is significantly different from zero or vice versa.

2.3.2. Applications of the Cox Test

The earliest applications of the Cox test involve testing of univariate probability distributions belonging to separate parametric families: see example 4 in chapter 1. Although Dhrymes et al. (1972) argued against the use of the Cox test for separate regression models, due to its perceived computational intractability, Pesaran (1974) derived this test and demonstrated the ability to compute the statistic with existing regression software. In the following classical linear regression models:

$$H_0: y = X\beta + u_0,\ u_0 \sim N(0,\ \sigma_0^2 I_n) \qquad (2.8)$$

$$H_1: y = Z\gamma + u_1,\ u_1 \sim N(0,\ \sigma_1^2 I_n) \qquad (2.9)$$

where H_0 and H_1 are non-nested by virtue of at least one distinct regressor in each of the matrices X and Z. Pesaran's Cox test for H_0 against H_1 is:

$$N_0 = [T_0 / (\hat{V}_0)^{1/2}], \qquad (2.10)$$

where $T_0 = (n / 2) \log (\hat{\sigma}_1^2 / \hat{\sigma}_{10}^2)$ and the denominator variance is,

$\hat{\sigma}_{10}^2 = \hat{\sigma}_0^2 + \hat{\beta}' X' M_z X \hat{\beta} / n$ whereas the denominator matrix of (2.10) is,

$$\hat{V}_0 = \hat{\sigma}_0^2 / \hat{\sigma}_{10}^4 M_z M_x M_z X \hat{\beta}$$
$$M_x = 1 - X (X'X)^{-1} X'$$
$$M_z = I - Z(Z'Z)^{-1} Z'$$

' ^ ' denotes the consistent estimators and n refers to the usual sample size.

An equivalent test statistic for H_1 against H_0 denoted as N_1 may be obtained by interchanging the roles of H_0 and H_1.

Although the statistics N_0 and N_1 seem to be cumbersome, but due to high speed computing power of recent days one can use either Guass or R language to do the computational work on PC. The following three steps in regression estimations are required.

(i) Run OLS of y on X and y on Z to obtain $\hat{\beta}$, $\hat{\sigma}_0^2$ and $\hat{\gamma}$, $\hat{\sigma}_1^2$, respectively. The vectors of predicted values, $\hat{y}_0 = X\hat{\beta}$ and $\hat{y}_1 = Z\hat{\gamma}$, should be saved for the next step.

(ii) Run OLS of $\hat{y}_0$ on Z and $\hat{y}_1$ on X to obtain $\hat{\sigma}_{10}^2$ and $\hat{\sigma}_{01}^2$ respectively. The residual vectors, $\hat{e}_0 = M_z\hat{y}_0$ and $\hat{e}_1 = M_x\hat{y}_1$, are also saved for the next step.

(iii) Run OLS of $\hat{e}_0$ on X and $\hat{e}_1$ on Z to obtain the residual sum of squares $\hat{\beta}' X' M_z M_x M_z X \hat{\beta}$ and $\hat{\gamma}' Z' M_x M_z M_x Z \hat{\gamma}$, respectively.

Pesaran and Deaton (1978) extend the Cox test to multivariate non-linear regression models where the test provides an added dimension of testing non-nested functional forms along with the selection of competing regressors. However, such test statistics are difficult to compute and require a series of full information maximum likelihood, non-linear regressions (see Pesaran and Deaton 1978, pg. 683-687). Note that in their illustrative application to consumption functions, Pesaran and Deaton demonstrate the lack of robustness of their test to autocorrelation.

The above tests are limited to a binary comparison of any two competing non-nested models. On the other hand, Sawyer (1984) extends the Cox test to handle the situation of multiple competing models in a joint test. The multiple tests may be more powerful than paired comparisons if all the competing models are well specified and sufficiently disparate (Sawyer 1984, pg. 424). Recently, Dastoor and McAleer (1987, pg. 65) conclude "that a joint test may be more widely consistent than a paired test."

Another extension of the Cox test is suggested by Aneuryn-Evans and Deaton (1980) to test linear versus logarithmic or additive versus multiplicative models. In an earlier study, however, Box and Cox (1964) used the Box-Cox transformations to evaluate such models. Asymptotically equivalent procedures based on Lagrange Multiplier principle are proposed by Andrews (1971) and Godfrey and Wickens (1981).

In spite of these generalizations, Cox's modified likelihood ratio approach and various extensions are not widely used in the empirical literature due to computational difficulties. The computationally simpler and asymptotically equivalent non-nested tests based on artificial embedding approaches are introduced in the next section.

2.4. Non-Nested Tests Based on Artificial Embedding

The pioneering concept of artificial embedding of competing hypotheses is credited to the early work of Hoel (1947). Non-nested hypothesis tests based on exponential and linear embedding are discussed in the following subsections.

2.4.1. Exponential Embedding

Atkinson (1970) amends Cox's test by constructing an exponential embedding for non-nested hypotheses. From the non-nested models H_0 and H_1 in (2.1) and (2.2), and denoting their likelihood functions as I_0 and I_1 respectively, Atkinson's exponentially-embedded density is:

$$I_I = \frac{I_0^{(1-\lambda)} I_1^{\lambda}}{\int I_0^{(1-\lambda)} I_1^{\lambda} dy} \tag{2.11}$$

where λ is the 'embedding' or 'nesting' parameter on which inferences are to be drawn, and the denominator is a normalizing constant. In logarithmic form, this density function may be expressed as:

$$L_\lambda = (1-\lambda)L_0 + \lambda L_1 - \log \left[\int I_0^{(1-\lambda)} I_1^{\lambda} dy \right] \tag{2.12}$$

where L_0 and L_1 denote the log likelihood functions under H_0 and H_1, respectively.

The truth of H_0 (or H_1) is ascertained by testing hypothesis $\lambda = 0$ (or $\lambda = 1$). In general, the log likelihood function L_λ is not estimable until one of the parameters, either λ or ß or γ, is identified. Therefore, to test $\lambda = 0$, Atkinson replaces γ with a specific value, say $\hat{\gamma}_{10}$, where $\hat{\gamma}_{10}$ denotes the asymptotic expectation under H_0, evaluated at $\beta = \hat{\beta}$. Then, using the Lagrange Multiplier principle, Atkinson obtains a test statistic for H_0 which seems to be similar to the Cox test. Pereira (1977) shows that, unlike the Cox test, Atkinson's test is not necessarily consistent although it is consistent for Gaussian models. Furthermore, Pesaran (1982) criticizes Atkinson's approach for being unable to generalize to multivariate non-nested econometric models.

2.4.2. Linear Embedding

Quandt (1974, pg. 93-94) discusses the difficulties in estimation of Atkinson's exponential density and prefers, instead, a linear embedding of the competing hypotheses. For non-nested, linear regression models such as (2.8) and (2.9), the linear embedding involves an artificial regression model given by:

$$H_A: y = (1-\lambda)\, X\beta + \lambda Z\gamma + u, \tag{2.13}$$

where λ, the nesting parameter, also acts as a testing parameter since a test of $\lambda = 0$ favors H_0 against H_1 . But λ is typically unidentified in the above regression and, therefore, no inference is possible.

However, to overcome on this identification problem, Davidson and MacKinnon (1981) propose replacing γ with $\hat{\gamma}$, the maximum likelihood estimator of γ, in (2.13):

$$H_J: y = (1-\lambda)\, X\beta + \lambda Z\hat{\gamma} + u. \tag{2.14}$$

Thus, a consistent and asymptotically standard normal test statistic, under H_0, may be obtained as a standard t-ratio for $\hat{\lambda}$. This is shown in Theorems 3.4 and 3.3.3 in next chapter.

A test based these theorems is called the J-test or simply J_0 test, as λ and ß are estimated jointly. It is expressed as follows:

$$J_0 = (\hat{\gamma}'Z'\, M_x\, y) / [\hat{\sigma}_J^2\, (\hat{\gamma}'Z'\, M_x\, Z\, \hat{\gamma})]^{1/2} \tag{2.15}$$

Where $\hat{\sigma}_J^2$ is the sum of squared residuals from H_J. The test for $\lambda = 1$ in (2.14) will not be a correct test for H_1; rather, a similar test is developed by reversing the roles of H_0 and H_1 in (2.14).

Davidson and MacKinnon also propose the asymptotically normal C- and P-tests. However, the C-test is a 'preliminary' test and does not have a correct size even when H_0 is true: see Theorem 3.3.4 in the next chapter. Other extensions of the P-test for nonlinear regression models, discussed by Davidson and MacKinnon (1982), can handle multivariate regressions as well as the additive versus multiplicative error structures.

The J-test under H_0 is shown to be asymptotically equivalent to the linearized Cox test of Pesaran (1974): see Theorem 3.3.2 of the next chapter.

The asymptotic equivalence of the P-test under H_0, with the Cox-type test of Pesaran and Deaton (1978), is also proved by Davidson and MacKinnon (1981). However, the ease of computation enhances the use of J- and P-tests in empirical applications.

In addition to the binary comparisons of the null and alternative hypotheses, these tests can be easily applied to the joint testing of multiple alternatives.

Fisher and McAleer (1981) develop an adjusted J-test, labelled the JA-test, by incorporating Atkinson's suggestion to replace in (2.14) with $\hat{\gamma}_{10}$:

$$H_{JA}: y = (1-\lambda)\, X\beta + \lambda Z\,\hat{\gamma}_{10} + u, \tag{2.16}$$

where $\hat{\gamma}_{10}$ denotes the asymptotic expectation of $\hat{\gamma}$ under H_0 , evaluated at $\beta = \hat{\beta}$. Again, a t-statistic is used to test the hypothesis $\lambda = 0$ for the truth of H_0 and test of H_1. This is done by obtaining the interchangeable roles of H_0 and H_1.

The JA-test is consistent, asymptotically normal under H_0, and asymptotically equivalent to the J- and Cox tests. However, small sample distributions of the J-, P-, and Cox-type tests are unknown. By using Milliken and Graybills (1970) results, Godfrey (1981, 1983) and McAleer (1981, 1983) show that JA-test follows an exact t-distribution with appropriate degrees of freedom. Godfrey and Pesaran (1983) propose a small sample adjustment to the Cox test, designed to correct for over-rejection of the true null in small samples. MacKinnon et al. (1983) also considers such an adjustment for the J-test but doubts its practical utility at the cost of complicating the otherwise simpler procedure.

Finally, it should be noted that many asymptotically valid tests can be constructed from the artificial regression model in (2.13), since γ can be replaced with any consistent estimator: see Pesaran (1982), Dastoor (1983), and Morimune (1983).

2.5. Tests Based on Instrumental Variables

The non-nested tests discussed above are based on maximum likelihood or least squares estimators of the competing regression models. Recognizing the need in economic modeling for instrumental variables (IV) estimation (arising, for instance, in the estimation of simultaneous equation systems or single equation, stochastic regressor models), Godfrey (1983), Ericsson (1983), MacKinnon et al. (1983), and Dastoor and McAleer (1985) propose IV-based non-nested testing procedures.

For the non-nested, linear regression models in (2.8) and (2.9), Godfrey's (1983) IV analogue of the Cox test is defined in the following. (The notation here conforms to the present study.)

$$d_0 = (1/n)\,[\tilde{u}'_1 A_w \tilde{u}'_i - \tilde{u}'_0 A_w \tilde{u}_0 - \tilde{\beta}'\hat{X}'\hat{M}_z\hat{X}\tilde{\beta}], \qquad (2.17)$$

where $\tilde{u}'_i$ is the vector of IV residuals from model i = 0, 1; $A_w = W(W'W)^{-1}W'$ and W is the IV matrix for both models; $\tilde{\beta}$ is the generalized IV estimator of β; = A_wX and = A_wZ the matrices of least squares predicted values obtained by regressing X on W and Z on W; $\hat{M}_z$ is the projection matrix, $I - (\hat{Z}'\hat{Z})^{-1}\hat{Z}'$. Godfrey (1983, pg. 360) showed that, under H_0, $\sqrt{n}\,d_0$ is asymptotically normal with zero mean and variance η_0^2; the consistent estimator of η_0^2 is given by:

$$\tilde{\eta}_0^2 = 4\,\tilde{\sigma}_0^2\,[\tilde{\beta}'\hat{X}'\hat{M}_x\tilde{X}\tilde{\beta}/n], \qquad (2.18)$$

where $\tilde{\sigma}_0^2 = (\tilde{u}_0'\tilde{u}_0/n)$, $\hat{X}' = \hat{M}_z\hat{X}$, and $\hat{M}_x = I - (\hat{X}'\hat{X})^{-1}\hat{X}'$. Under H_0, an asymptotic standard normal test statistic, $G_0 = \sqrt{n}\,d_0/\hat{\eta}_0$, is formally named the G-test by Godfrey.

Another test of Godfrey, is named as the T-test, which is obtained in an analogue of the JA-test from IV estimation of following artificial regression model:

$$H_T\colon y = X\beta + X\beta\lambda + u.$$

The t-statistic for λ is obtained to test the truth of H_0. It is shown by Godfrey that both the T- and G-tests under H_0 are asymptotically equivalent; however, the computation of T is much simpler than G. The T-test, like the JA-test, follows an exact t-distribution with appropriate degrees of freedom if X and Z include all non-stochastic regressors. Note that, in such cases, the IV estimators are equivalent to the OLS estimators and the T-test also equals the JA-test in value but differs in sign.

Ericsson (1983) also proposes various IV statistics for nested and non-nested models. His Cox-type test is numerically equal to Godfrey's G-test, while another is obtained as the Wald encompassing test of Mizon and Richard

(1986) from IV estimation of a comprehensive model such as (2.4). Ericsson compares these tests under local and comprehensive alternatives but is unable to determine any uniform ranking of their asymptotic powers.

MacKinnon, White, and Davidson (1983) modify the J-test by replacing the OLS estimators in (2.14) with 2SLS estimators and obtain a computationally convenient test for non-nested linear models where some of the regressors are endogenous. By imposing certain regularity conditions, they prove that, under H_0, their statistic is asymptotically standard normal.

Dastoor and McAleer (1985) generalize the R procedure of Dastoor (1983) and consider the generalized IV estimation of the comprehensive model in (2.4) resulting in the standard F-test. Such a statistic is asymptotically equivalent to the Wald test of Ericsson (1983) and, under H_0, it is distributed as an asymptotic X_r^2 where r is the number of distinct regressors in H_1. Computational simplicity is considered to be its attractive feature in practical applications; however, lower asymptotic power than the 'one-degree-of-freedom' tests such as the J-test under Pesaran type local alternative may be a serious drawback.

2.6. Encompassing Principle

The encompassing principle is concerned with the ability of a model to account for the behaviour of others, or less ambitiously, to explain the behaviour of relevant characteristics of other models. (Mizon 1984, pg. 136)

In the case of nested models, a general model always encompasses the restricted models. But for non-nested models, Mizon (1984, pg. 137) notes that "... highly parametrized models are likely to encompass parsimoniously parameterized ones."

For the non-nested, linear regression models in (2.3) and (2.4), the formal definition of Mizon and Richard (1986, pg. 660) is invoked in the following way which help to develop Wald Encompassing Test (WET).

DEFINITION

H_0 encompasses H_1 with respect to $\hat{\gamma}$ if $\hat{\phi} = \hat{\gamma} - \hat{\gamma}_{10}$ does not significantly differ from zero.

Thus, the WET statistic for from (2.4) is given by Mizon (1984, pg. 149) as:

$$\eta_w(\hat{\gamma}) = y' M_x (\overline{Z}' M_x \overline{Z})^{-1} \overline{Z}' M_x y / \hat{\sigma}_0^2, \qquad (2.19)$$

where $\overline{Z}$, consists of the r distinct columns of Z. It is shown that, under H_0, $\eta_w(\hat{\gamma})$ is asymptotically distributed as χ^2 with r degrees of freedom and is asymptotically equivalent to the conventional F-test of $\gamma = 0$ in comprehensive model (2.5). Note that the R-procedure of Dastoor (1983) can also be interpreted as a WET statistic.

Analogously, the variance encompassing test for $\hat{\sigma}_1^2$ is defined by Mizon (1984, pg.150) to be:

$$\eta_W(\hat{\sigma}_1^2) = \frac{[(n-r)(\hat{\sigma}_1^2 - \hat{\sigma}_0^2) - \beta' X' M_Z X \beta]^2}{4\hat{\sigma}^2 \hat{\beta}' X' M_Z M_X M_Z X \hat{\beta}}$$

following an asymptotic x_1^2 under H_0 . Mizon and Richard (1986, pg. 667-669) also propose a Complete Parametric Encompassing (CPE) test with respect to both $\hat{\gamma}$ and $\hat{\sigma}_1^2$. Note that the expression for the CPE test is the same as the WET in (2.19).

Mizon and Richard (1986) prove the asymptotic equivalence between $\eta_w(\hat{\sigma}_1^2)$ and non-nested tests such as J-, JA-, Pesarans version of Cox test and Sawyer's (1983) test based on an information criterion. Mizon and Richard (1986) further extend the proof of the asymptotic equivalence between variance encompassing test statistic and non-nested tests of Godfrey (1983). Thus, they emphasize the usefulness of these encompassing principles for unifying the various approaches of testing non-nested models and argue that it "should prove a powerful tool in empirical modeling." However, Mizon and Richard outline several important unresolved issues concerning the application of encompassing tests:

(i) The robustness of encompassing procedures under weak erogeneity assumptions, and

(ii) [Construction of] large sample approximations to [γ_{10}], and more generally of applying Monte Carlo simulation procedures to evaluate encompassing test statistics. (Mizon and Richard 1986, pg. 674)

Both of these issues will be addressed in this book. The presence of errors-in-regressors causes a failure of the erogeneity assumption; therefore, the large as well as small sample behavior of non-nested tests will be studied under such situations.

2.7. EMPIRICAL APPLICATIONS

MacKinnon (1983a) portrays a gloomy picture of the empirical work on non-nested hypothesis testing. The following list, however, provides evidence of a growing interest by applied econometricians in non-nested testing procedures.

Studies investigating non-nested consumption functions may be found in Pesaran and Deaton (1978), Sawyer (1980), Davidson and MacKinnon (1981), and Fisher and McAleer (1981). Separate demand systems are studied by Deaton (1978) and Davidson and MacKinnon (1983). Comparisons of non-nested money demand functions are carried out by McAleer, Fisher and Volker (1982), Gregory and McAleer (1983), Smith and Maddala (1983), Johannes and Nasseh (1984) and Thornton (1985). Bean (1981), Wisely and Johnson (1985) and Bemanke et al. (1988) study separate investment functions. Pesaran (1982a) and Dadkhah and Valbuena (1985) test Barro's model of unemployment against the Keynesian natural rate of unemployment.

Harvey (1977), Beath (1979), Volker (1980), Green, Pope and Phipps (1981), McAleer and Fisher (1982) and Pagan, Hall and Trivedi (1983) also provide interesting applications of the non-nested hypothesis testing to various economic phenomena.

It is interesting to note that most of these studies have resulted in joint acceptance or rejection of the studied models.

One possible explanation of this outcome is that non-nested tests are model specification tests, and if the competing models fail any percussive diagnostic checks then they have little chance of surviving a non-nested analysis.

Also, the observed over rejection of both models in many of these studies may well be attributed to small-sample distortions of the sampling distributions of these asymptotic tests, a topic discussed in the following section.

2.8. Statistical Properties

In general, it is possible to characterize the asymptotic behavior of the non-nested tests analytically; the small sample properties are typically obtained through specific Monte Carlo simulation experiments. Even the exact tests (the F-test, JA-test, R-test, and T-test) have various undesirable statistical properties - most notably, low power. This section summarizes some of the asymptotic properties and Monte Carlo results that have arisen in the literature.

2.8.1. Asymptotic Distributions and Size of the Tests

All non-nested tests, except the F-test and its equivalent tests based on the R-procedures, are termed "one-degree-of-freedom" tests, since each is asymptotically standard normal, whose square then follows an asymptotic X_1^2 when H_0 is true. The noted exceptions are also asymptotically χ^2, although the degrees of freedom will depend on the number of restrictions imposed on the associated comprehensive model.

For instance, in the choice of non-overlapping regressor sets, the degree of freedom for the test of H_0 from the comprehensive model in (2.4) is the number of distinct regressors in H_1.

Pesaran (1982, pg. 269) provides the following definition: "by the 'size' of a test is usually meant the magnitude of its Type I error." Thus, all non-nested tests which have correct asymptotic distributions will also have the correct size asymptotically.

The C-test of Davidson and MacKinnon is a notable exception, possessing a Type I error, which is less than the nominal size of the test. It should be noted, however, that there is a well-documented tendency of the non-exact tests towards highly distorted Type I errors in small samples. These results are discussed in more detail in chapter three.

2.8.2. Consistency of the Tests

A test is consistent if the probability of rejecting a false null hypothesis approaches one asymptotically.

The proof of consistency involves the examination of the probability limit of the null statistic under the alternative. For example, since the Cox test is zero mean under H_0, if it can be shown that $T_0 \neq 0$, this implies that H_0 will be rejected with unit probability if H_1 is true. See Pereira (1977), Pesaran (1982), Davidson and MacKinnon (1982) and Ericsson (1983) for formal proofs of the consistency of various non-nested tests. Regarding the Cox and J-tests, the proof of consistency will be outlined in the next chapter.

2.8.3. Asymptotic Power of the Tests

The probability of making correct decisions determines the power of a non-nested test. However, a uniform ranking of non-nested tests in terms of their powers has been a challenging task. Three distinct approaches are suggested in the literature to compare the asymptotic power of consistent tests.

(i) Pesaran (1982) considers the "local alternative approach" where H_1 approaches H_0 asymptotically and the task of distinguishing between the two becomes increasingly difficult as the sample size gets larger and larger. Pesaran shows that the Cox and J-tests, under local alternatives, have the same asymptotic power. Since both are distributed as a non-central X_1^2 and the non-centrality parameter is the same for both tests (formal proofs are outlined in chapter three). However, the power of the Cox and J-tests exceed that of the F-test when the number of non-overlapping variables, r, in H_1 exceeds 1 since the F-test, following a X_r^2, is then dominated by the X_1^2X tests. This result follows from Das Gupta and Perlman (1974). When r = 1, all three tests will have the same asymptotic power. Ericsson (1983) derives asymptotic power estimates of IV-based non-nested tests under local alternatives; a conclusive ranking of his tests could not be obtained.

(ii) Davidson and MacKinnon (1982) propose the "fixed alternative" approach where H_1 remains fixed but the data generating process (DGP) collapses to H_0 asymptotically. This approach is useful when neither H_0 nor H_1 is true. A formal derivation of powers under the fixed alternative approach will be presented in the next chapter.

(iii) Pesaran (1984) considers Bahadur's (1960, 1967) approach of comparing the approximate slopes of separate tests. In this case, H_1 remains fixed but the size of the test approaches zero in large samples. The approximate slope of a test is derived as a probability limit under

H_1 of the squared value of null statistic, e.g., $\text{plim}_{H_1}(J^2_0/n)$ represents the approximate slope of the J-test. The greater the value of the approximate slope of a test, the greater will be its asymptotic power. Pesaran considers a specific linearized version of Cox test and demonstrates its asymptotic efficiency over the F and J-tests. In the case of Davidson and MacKinnon's (1981a) linearized form of Cox test, it will be shown in the next chapter that such relative efficiency over the J-test does not hold asymptotically.

2.8.4. Monte Carlo Evidence of Small Sample Properties

In practical applications, small sample behavior of various test statistics is an important issue. The limited evidence of small sample sizes of non-nested tests is summarized from Pesaran (1974), Pesaran (1982a), Davidson and MacKinnon (1982) and Godfrey and Pesaran (1983).

Although the results of Monte Carlo experiments are design specific and may not be generalized to different situations, important information has been obtained from these studies.

(i) Pesaran (1974) shows that in samples of size 40 or less, the nominal size of the Cox test is overstated by 3 percent while the F-test has the same empirical and nominal size. In terms of power, however, the F-test is dominated by the Cox test and the superiority of the latter increases with the number of non-overlapping variables in H_0 and H_1: see Pesaran (1982).

(ii) Godfrey and Pesaran (1983) observe similar small sample over-rejection in the range of 15 - 25 percent by the J-test, although comparisons of the Cox and J-tests reveal that small sample size of the J-test is usually greater than the Cox test. The factors contributing to the size problem are poor fit of H_0, low colinearity between the regressors of H_0 and H_1, and a false model including more regressors than the true model. The proposed small sample adjustment of the Cox test is shown to be effective in reducing the size of the test, although these adjustments cannot be easily computed using existing software.

(iii) The simulation experiments of the above authors support the earlier claim that the JA-test has an exact small sample distribution, as well as a very low power relative to the adjusted Cox test and the F-test when the number of regressors of the "true" DGP in H_0 exceeds that in H_1. This lack of power increases with increasing canonical correlations between the regressors of competing models.

(iv) Davidson and MacKinnon (1982) demonstrate that the JA- and J-tests have equal power when testing against the true model, but the JA-test has less power when testing against false models. However, McAleer (1985) notes that it is not clear whether the low power of the JA-test is due to testing against the false alternatives or difference in the number of the regressors in the competing models.

2.9. CONCLUSION

Cox (1962, pg. 406) summarizes the intended purpose of non-nested tests: "high power is desired against a composite alternative hypothesis that is not in the same parametric family as the null hypothesis." To this end, Cox's criterion is typically considered to be the most powerful approach, but computational difficulties may limit its appeal for real life practical applications. Even in today's high computing power, real life applications of these tests are difficult (see Hong and Preston (2011).

On the other hand, the tests based on linearly-embedded, artificial regression models accommodate ease of application; distorted Type I errors and/or low power in small samples are their major weaknesses. Monte Carlo evidence to this effect is primarily limited to linear regression models, while little evidence appears on the small sample performance of non-linear model tests such as the C- and P-tests. The comparison of the multiple and binary tests and testing the robustness of existing non-nested tests are also unresolved issues.

In general, the existing non-nested tests deal with separate classical regression models under full ideal conditions. On the other hand, empirical problems do not typically support the absence of serial correlation or heteroscedasticity, non-stochastic regressors, and so forth. For instance, McAleer et al. (1982) show that the presence of irrelevant variables in an alternative model does not cause inconsistency for the test of null model; however, incorrect omission of variables may render a non-nested test to be inconsistent.

In this context, the present book aims at evaluating various non-nested tests when mismeasured or unobserved regressors are present in separate regression models. Although the IV-based non-nested tests are consistent, the issues of asymptotic efficiency and small sample comparison of power and significance levels of these tests will be examined in this book. To begin, let's consider asymptotic theory of non-nested tests in the next chapter.

Chapter 3

ASYMPTOTIC THEORY AND THE NON-NESTED TESTS

3.1. INTRODUCTION

Recently, Hong and Preston (2011) discussed in great detail the importance of the asymptotic theory of no-nested testing in regression. As asymptotic theory has special significance in the context of errors-in-variables models (EVM) as well as in non-nested testing. As noted earlier in previous chapters, the property of consistency for EVM estimates is usually possible in large samples only when an asymptotically valid instrumental variables (IV) estimation procedure is employed.

In the case of non-nested tests, it has also been emphasized that most of the separate testing procedures are only asymptotically valid. Thus, a formal introduction to asymptotic theory, with particular emphasis on the EVM and non-nested tests, is included in this chapter.

The structure of this chapter is as follows. The next section introduces the fundamental concepts of asymptotic theory. It provides formal basic definitions of the concepts detailed in the proceeding sections and the chapters and some needed notations are given for the ready reference of the reader.

In section 3.3, the case of stochastic regressors as a precursor to the EVM is presented to demonstrate the inconsistency of OLS estimators and the consistency of the new IV estimators.

Asymptotic distributions theory related to non-nesting, its consistency, and the concept of asymptotic power of various non-nested test statistics are derived in section 3.4. This section presents linearized Cox test, J-Test and a size and power function of these tests are discussed. Section 3.5 concentrates on discussing consistency of non-nested tests and section 3.6 presents asymptotic power of various tests under alternative which include Cox test, J-Test and an F-test. The final section concludes with the brief ending remarks.

3.2. Fundamental Concepts of Asymptotic Theory

Several concepts such as limits, convergence, orders of magnitudes and their stochastic counterparts are given here using formal definitions and notation in the following subsections.

3.2.1. Basics; Limits, Convergence of Vectors, and Matrices

Definition 1

Consider a sequence of n real numbers, denoted as $\{a_n\}$. Then a finite real number 'a' may be defined as the *limit* of $\{a_n\}$ if $|a_n - a| < s$ for all $n \geq N$, where s represents a positive real number and N is an integer

Definition 2

If the sequence $\{a_n\}$ has a limit 'a' then $\{a_n\}$ is said to be *convergent* at 'a' as n approaches infinity, formally written as: if $\lim_{n\to\infty} a_n = a$ then $a_n \to a$. On the other hand, if $a_n \to \infty$ then a_n will be termed as *a divergent* sequence.

Definition 3

Let a_n now denote a row or column vector of k elements, such as a_{n_i} for $i = 1,2, \ldots k$. If $a_{n_i} \to a_i$, for all i, then $a_n \to a$ where 'a' is a k dimensional vector. Similarly, if a represents a matrix of order kxr and each element of the matrix has a corresponding finite limit in some matrix 'a' of the same order as a_n then again $a_n \to a$.

Note that if $a_n \to a$ and $b_n \to b$ then following results hold whether a_n and b_n are sequences, vectors, or matrices:

(i) $a_n + b_n \to a + b$
(ii) $a_n \,.\, b_n \to a \,.\, b$

3.2.2. Order of Magnitudes

The concepts of order of magnitudes are closely related with the concepts of limits, illustrated within the following definitions:

DEFINITION 4

The sequence $\{a_n\}$ is said to be *at most of order* n^r and is denoted as $a_n = 0(n^r)$ if there exists a finite real number 'a' such that $n^{-r} |a_n| < a$. Note that if $a_n = 0(n^0)$ or, equivalently, $a_n = 0(1)$ then $\{a_n\}$ is called a *bounded* sequence.

DEFINITION 5

The sequence $\{a_n\}$ is of *smaller order* than n^r and is denoted as $a_n = o(n^r)$ if $\lim_{n\to\infty} n^{-r} |a_n| = 0$.

Similar to the definitions of limits, the definitions of orders of magnitudes may be extended to vectors and matrices if each element of a vector or matrix satisfies the above definitions.

Some major results concerning the orders of magnitudes, for $a_n = o(n^r)$ and $b_n = o(n^s)$, may be stated as follows:

1. $a_n + b_n = O(\max[n^r, n^s])$
2. $a_n . b_n = O(n^{r+s})$
3. $a_n^s = O(n^{rs})$ for $s > 0$.

Similar results will hold for smaller order if O is replaced with o. However, if $a_n = 0(n^r)$ and $b_n = o(n^s)$ then $a_n . b_n = o(n^{r+s})$. White (1984) provides the proofs for these and several other propositions regarding orders of magnitudes.

3.2.3. Stochastic Limits, Convergence, and Orders

This section consists of the stochastic counterparts of the above results. Mann and Wald (1943) provided an earlier effort to organize and systematically present the definitions, notation, and propositions regarding probability limits and order relationships. Subsequent discussions of these concepts are available in Rao (1973), Fuller (1976), White (1984), and Judge et al. (1986). The following definitions are important for formalizing the concepts used in this book.

DEFINITION 6

Let $\{a_n\}$ now represent a sequence of random variables. The sequence $\{a_n\}$ converges in probability to a random variable 'a' if for every positive values, $\lim_{n\to\infty} P[|a_n - a| < s] = 1$. Thus, the random variable 'a' is referred as the *probability limit* of $\{a_n\}$ and is denoted as plim $a_n = a$ or $a_n \xrightarrow{P} a$.

Analogously, if $\{a_n\}$ represents a sequence of random vectors or matrices rather than the scalar random variables, then 'a' will be a vector or matrix respectively.

DEFINITION 7

A stronger concept of convergence in probability than the one defined above, termed *almost sure convergence*, is defined as:

if $P [\lim_{n\to\infty} |a_n - a| < s] = 1$ |a then $a_n \to a$.

The two concepts of convergence are differentiated as follows:

With almost sure convergence the probability measure P takes into account the joint distribution of the entire sequence $\{a_n\}$, [notation conforms with the present study], but with convergence in probability, we only concern ourselves sequentially with the joint distribution of elements of $\{a_n\}$. (White 1984, pg. 22)

In most econometric applications, simple convergence in probability is invoked; therefore, further details of its comparison with almost sure convergence are skipped here.

A very important result regarding econometric applications of probability limits, referred as Slutsky's theorem, is stated by Judge et al. (1986, pg. 147) in the following words: "if g(.) is a continuous (vector) function, then plim $[g(a_n)] = g[\text{plim } a_n]$."

Analogous to orders of magnitudes, there exist definitions of orders in probability such as:

DEFINITION 8

The sequence $\{a_n\}$ of random variables is *at most of order n^r in probability,* if for all n and $s > 0$ there exists a real number N such that $P [n^{-r}$

$|a_n| \geq N] < s$ and it is denoted as $a_n = O_p(n^r)$. When $a_n = 0_p(n^\circ)$ or, equivalently, $a_n = 0_p(1)$, a_n is said to be bounded.

DEFINITION 9

The *smaller order in probability* of the sequence $\{a_n\}$ is defined as n^r if Plim $n^{-r} a_n = 0$ and will be written as $a_n = 0_p(n^r)$.

Once again, these definitions are workable if a_n denotes the sequence of random vectors or matrices. Furthermore, the results stated for orders of magnitudes are also valid for orders in probability.

3.3. Consistency of Estimators and Stochastic Regressors

The case of stochastic regressors demonstrates the application of probability limits in determining the consistency of econometric estimators. To begin, let's consider the case of standard classical regression model of section 1.1 which is summarized below:

$$y = X\beta + u.$$

Then the least square estimators of the parameter vector β are defined as:

$$\hat{\beta} = (X'X)^{-1} X'y = \beta + (X'X)^{-1} X'u.$$

Suppose that $\lim_{n\to\infty} \frac{X'X}{n} = Q_{xx}$ and plim $\frac{X'u}{n} = 0$, where Q_{xx} is a positive definite matrix of order kxk and 0 is kxl vector of zeros, then it can be easily shown with the help of Slutsky's theorem that:

$$\text{plim } \hat{\beta} = \beta + \lim_{n\to\infty}\left(\frac{X'X}{n}\right)^{-1} . \text{ plim } \frac{X'u}{n}$$

$$= \beta + Q_{xx}^{-1} . 0 = \beta,$$

which implies that $\hat{\beta}$ is a *consistent* estimator of β. A formal definition of consistency may be given in definition 10 below.

DEFINITION 10

If a random vector of estimators, $\hat{\beta}$, converges in probability to its parameter vector, β, then $\hat{\beta}$ is said to be consistent estimator of β.

Next, consider the case where matrix X in the above model involves one or more stochastic regressors such that plim $\frac{X'u}{n} \neq 0$. Then from the earlier proof it follows that plim $\hat{\beta} \neq \beta$. Thus, $\hat{\beta}$, the OLS estimator, is inconsistent. The EVM may be considered as a special case of the stochastic regressors problem where plim $\frac{X'u}{n} \neq 0$. Thus, the derivation of consistent estimators under such situations is usually carried out with the help of IV estimation.

An nxl matrix W, with $1 \geq k$, may be defined as an IV matrix for X if it satisfies the following assumptions:

(i) plim (W'X/n)= Q_{wx} where Q_{wx} is finite and non-singular
(ii) plim (W'W /n)= Q_{ww} , where Q_{ww} is finite, positive definite and non-singular
(iii) plim (W'u/n)= 0 where 0 is Ix1 vector of zeros.

The generalized IV estimator of β is defined by Bowden and Turkington (1984, pg. 13) as:

$$\tilde{\beta} = [X'\, A_w^{-1} X]^{-1} A_w\, y = \beta + [X'\, A_w^{-1} X\,]^{-1}\, X' A_w\, u, \qquad (3.1)$$

where $A_w = W(W'W)^{-1}\, W'$. Thus, the consistency of $\tilde{\beta}$ follows:

plim $\tilde{\beta} = \beta + [Q_{xw}\, Q_{xw}^{-1}\, Q_{wx}]^{-1}\, Q_{xw}\, Q_{ww}^{-1}\, .0 = \beta$

3.4. Asymptotic Properties of Non-Nested Tests

This section deals with the fact that, when the null hypothesis is true, most of the non-nested test statistics have asymptotically well-defined distributions and correct sizes. Following Davidson and MacKinnon (1981a), asymptotic distributions of the linearized Cox test and J-test, under the null hypothesis, are re-derived to illustrate their asymptotic equivalence. It is also demonstrated that the C-test of Davidson and MacKinnon (1981a) does not have the correct asymptotic size under the null hypothesis. Although the major emphasis here is on asymptotic distributions of non-nested tests, the exact JA-test of Fisher and McAleer (1981) is also briefly discussed.

Consider the non-nested linear regression models of equations (2.8) and (2.9):

$$H_0: y = X\beta + u_0, \quad u_0 \sim N(0, \sigma_0^2 I_n) \tag{3.2}$$

$$H_1: y = z\gamma + u_1, \quad u_1 \sim N(0, \sigma_1^2 I_n) \tag{3.3}$$

Both models are assumed to satisfy the standard classical assumptions.

Thus, $\lim_{n\to\infty}(\frac{X'X}{n}) = Q_{xx}$ and $\lim_{n\to\infty}(\frac{Z'Z}{n}) = Q_{zz}$ are non-singular and positive definite. Also, assume that matrices X and Z are non-orthogonal, formally, $\lim_{n\to\infty}(\frac{X'Z}{n}) = Q_{xz} = Q'_{zx}$ is a finite and non-zero matrix.

To test H_0 against H_1, the Cox-type test of Pesaran (1974) and the J-test of Davidson and MacKinnon (1981a), denoted as N_0 and J_0 respectively, may be defined as:

$$N_0 = [T_0 / \hat{V}_0^{1/2}] \tag{3.4}$$

$$J_0 = [\hat{\lambda}_J / (\hat{\lambda}_J)^{1/2}] \tag{3.5}$$

where

$$T_0 = (n/2) \log(\hat{\sigma}_1^2 / \hat{\sigma}_{10}^2) \tag{3.6}$$

$$\hat{V}_0 = (\hat{\sigma}_0^2 / \hat{\sigma}^4{}_{10}) \hat{\beta}' M_z M_x M_z X \hat{\beta} \tag{3.7}$$

$$\hat{\sigma}_{10}^2 = \hat{\sigma}_0^2 + (1/n) \hat{\beta}' X' M_z X \hat{\beta}$$

$$\hat{\sigma}_0^2 = (y' M_x y / n) \tag{3.8}$$

$$\hat{\lambda}_J = [y' A_z M_x y / y' A_z M_x A_z y] \tag{3.9}$$

$$\hat{V}\,(\hat{\lambda}_J) = \hat{\sigma}_J^2\,[\,y'\,A_z M_x\,A_z y\,]^{-1} \tag{3.10}$$

$$\hat{\sigma}_J^2 = \frac{y'M_x y}{n} - \frac{(y'A_z M_x y)^2}{n\,(y'\,A_z\,M_x\,A_z\,y)} \tag{3.11}$$

In (3.11), the matrices are M are;
$M_x = I - A_x,\ A_x = X(X'X)^{-1}X'$
$M_z = I - A_z,\ A_z = Z(Z'Z)^{-1}Z'$

Where the notation ' ^ ' denotes maximum likelihood estimators and n refers to the usual sample size.

The following theorems are presented to formalize the asymptotic structure of these test statistics.

THEOREM 3.3.1

Under H_0, T_0 in (3.6) is $O_p(n^{1/2})$ and $\hat{\lambda}_J$ in (3.9) is $O_p(n^{-1/2})$.

PROOF

From (3.7) above, the variance of T_0 under H_0 can be written as:

$$V_0 = (\sigma_0^2/\sigma_{10}^4)\,\beta'X'\,M_z M_x M_z X\beta. \tag{3.12}$$

Since and are assumed to be finite constants, and $\beta'X'\,M_z\,M_x\,M_z\,X\beta$ is O(n), a quantity that exists by virtue of the presumed existence of Q_{XX}, Q_{ZZ}, and Q_{ZX}. Thus, V_0 in (3.12) is also O(n). Hence, T_0 is $O_p(n^{1/2})$.

Similarly, from (3.10) the variance of $\hat{\lambda}_J$ under H_0 is given by:

$$V(\hat{\lambda}_J) = [\beta'X'\,A_z\,M_x\,A_z\,X\,\beta]^{-1} \tag{3.13}$$

Following the earlier arguments, it can be readily seen that $V(\hat{\lambda}_J)$ is $0(n^{-1})$.

Thus, $\hat{\lambda}_J$ is $O_p(n^{-1/2})$.

COROLLARY 3.3.1

The statistics $T_0 / \sqrt{n}$ and $\sqrt{n}\ \hat{\lambda}$ are O_p (1) and each possesses a well-defined limiting distribution.

PROOF of the above corollary follows directly from above theorem.

The import of this corollary is that the remainder of this study will deal with tests written in O (1); for example, the Cox and J-tests in (3.4) and (3.5) can be expressed as:

$$N_0 = [(T_0/\sqrt{n})/(\hat{V}_0/n)^{1/2}] \tag{3.14}$$

$$J_0 = [(\sqrt{n}\ \hat{\lambda}_J)/(n\ \hat{V}(\hat{\lambda}_J))^{1/2}] \tag{3.15}$$

The following material introduces linearized forms of the Cox test, approximations used frequently in the literature, since comparisons of the Cox test (in its original form) with other non-nested tests are difficult due to the logarithmic expression in (3.6): for example, see Davidson and MacKinnon (1981a), White (1982a), and Pesaran (1984). The linearized form suggested by Davidson and MacKinnon (1981a) will be used below.

3.4.1. Linearized Cox Test

THEOREM 3.3.2

Pesaran's Cox-type statistic, $T_0 \sqrt{n}$ can be rewritten in "linearized form" as:

$$S_0 = -(\sqrt{n}/2)\,[(\hat{\sigma}_{10}^2 - \hat{\sigma}_1^2)/\hat{\sigma}_1^2], \tag{3.16}$$

and an asymptotic standard normal statistic under H_0, denoted N_{L_0}, is given by:

$$N_{L_0} = [S_0\ /\ (\hat{V}_0/n)^{1/2}] \tag{3.17}$$

PROOF

The proof of this theorem is due to Davidson and MacKinnon 1981a, pg. 789. It is important to note that the first part of this proof requires to show the equivalence between $(T_0/\sqrt{n})$ and S_0.

Note that $T_0/\sqrt{n} = (\sqrt{n}/2) \log (1\text{-}2S_0\sqrt{n})$. Then, by applying a Taylor series expansion formula for $\log (1\text{-}2S_0/\sqrt{n})$, we get

$$T_0/\sqrt{n} = S_0 + 2S_0^2/\sqrt{n} + 4S_0^3/n + \dots$$

When $n \to \infty$, the terms of lower order of magnitude will disappear, implying that $(T_0/\sqrt{n} \to S_0$. Thus, S_0 is an asymptotically equivalent form of $T_0/\sqrt{n}$, completing the first part of the proof.)

Turning to the asymptotic normality of S_0, after substitution, $X\beta + u_0$ for y in S_0 above and with some manipulation, the expression for S_0 may be rewritten as:

$$S_0 = [(\beta'X' M_z M_x u_0/\sqrt{n})/\hat{\sigma}_{10}^2], \tag{3.18}$$

where $\hat{\sigma}_{10}^2 = \hat{\sigma}_0^2 + \beta'X' M_z X\beta/n$. Recalling that $u_0 \sim N(0, \sigma_0^2 I_n)$ then plim $(S_0) = 0$ is obvious, and the variance of S_0 is given by the expression:

$$V_0 = (\hat{\sigma}_0^2/\hat{\sigma}_{10}^4)(\beta'X' M_z M_x M_z X\beta/n),$$

from direct following equation (3.18). Then, by the Lindberg-Feller central limit theorem, the distribution of $S_0 \sim N(0, V_0)$ asymptotically.

The consistent estimator of V_0 can be written in the following format:

$$(\hat{V}_0/n) = (\hat{\sigma}_0^2/\hat{\sigma}_{10}^4)(\hat{\beta}'X' M_z M_x M_z X/n),$$

which is exactly the variance estimator of $(T_0/\sqrt{n})$. Thus, an asymptotic standard normal statistic for H_0, termed the “linearized. Cox test” and is denoted as N_{L_0}, can be expressed by (3.17).

3.4.2. J-Test

THEOREM 3.3.3

The J-test, obtained as a t-statistic for from the artificial regression model H_J: $y = X\beta (1 - \lambda_J) + \lambda_J Z\hat{\gamma} + u_J$, is asymptotically distributed as N(0, 1) when H_0 is true.

PROOF

Using Davidson and MacKinnon 1981a, "LEMMA 1", pg. 786, the consistent estimators of λ_J which denoted as $\hat{\lambda}_J$ and the variance of λ_J, denoted as $\hat{V}(\hat{\lambda}_J)$, is obtained from least squares estimation of H_J. Both estimators are given below:

$$\hat{\lambda}_J = [y' A_z M_X y / y' A_z M_X A_z y]$$

$$\hat{V}(\hat{\lambda}_J) = [y' A_z M_x A_z y]^{-1}$$

where $\hat{\sigma}_J^2$, estimated variance of H_J is given by:

$$\hat{\sigma}_J^2 = [y' M_x y]/n - \{[y' A_z M_x A_z y]^2\} / n[y' A_z M_x A_z y]$$

Following Corollary 3.3.1, when H_0 is true, the statistics under the null hypothesis is expressed as:

$$\sqrt{n}\,\hat{\lambda} = \frac{[u_0' A_z M_x u_0 + \beta' X' A_z M_x u_0] / \sqrt{n}}{[\beta' X' A_z M_x A_z X\beta + 2\beta' X' A_z M_x u_0 + u_0' A_z M_x A_z u_0] / n}$$

Now, retaining terms of the highest order in probability, the above expression under null can be simplified below:

$$\sqrt{n}\,\hat{\lambda} = \frac{\beta' X' A_z M_x u_0 / \sqrt{n}}{\beta' X' A_z M_x A_z X\beta / n} \quad (3.19)$$

Again, since $u_0 \sim N(0, \sigma_0^2 I_n)$, then, following the logic of Theorem 3.3.2, it can be easily shown that the distribution of statistics given in equation (3.19) is normal with zero mean and variance as $\sim N(0, n\, V(\hat{\lambda}_J))$ where $V(\hat{\lambda}_J) = [\beta' X' A_z M_x A_z X\beta]^{-1}$

Note that the estimated value $V(\hat{\lambda}_J)$ is a consistent estimator of $V(\hat{\lambda}_J)$, since $\text{plim}_{H_0}\, \hat{\sigma}_J^2 = \sigma_0^2$ and $\text{plim}_{H_0} (y' A_z M_x A_z y / n) = \lim_{n\to\infty} [\beta' X' A_z M_x A_z X\beta/n]$. Thus, an asymptotically standard normal statistic under H_0, named the J-test, may be defined as:

$$J_0 = \frac{\sqrt{n}\, y' A_z M_x y}{[\hat{\sigma}_J^2 (y' A_z M_x A_z y / n)]^{1/2}} \tag{3.20}$$

It should be noted that the truth of H_1 can be construed by constructing new test statistics,

N_{L_1} and J_1, whose values are inferred from (3.17) and (3.20) respectively, after interchanging the roles of H_0 and H_1 .

3.4.3. Asymptotic Equivalence of Linearized Cox Test and J-Test

THEOREM 3.3.4

The linearized Cox test and J-test are asymptotically equivalent under H_0 but have opposite signs.

PROOF

The proof of the above theorem may be obtained by using Davidson and MacKinnon 1981a "LEMMA 3", pg. 789.

The comparison of equations (3.18) and (3.19), representing numerators of linearized Cox-test and J-test under H_0, results in a simple correlation coefficient of -1. Moreover, the variances for both expressions converge to the same probability limits. Thus, when H_0 is a true model, the linearized Cox-test and J-test have asymptotically equivalent expressions with opposite signs.

This asymptotic result surfaces in small samples as well, as demonstrated by the empirical comparison of these tests: see, e.g., Johannes and Nasseh (1985). Since these tests are two tailed, their opposing signs will not alter the inference. Furthermore, the asymptotic equivalence of these tests cannot be established under H_1 since both tests do not converge to the same values, in particular:

$$\operatorname{plim}_{H_1} S_0 = - (\sqrt{n} /2) \ \operatorname{plim}_{H_1} [\hat{\sigma}_{10}^2 - \hat{\sigma}_1^2) / \hat{\sigma}_1^2]$$

$$= \frac{-\sqrt{n}\gamma'(Q_{zz} - Q_{zx} Q_{xx}^{-1} Q_{xz} Q_{zz}^{-1} Q_{zx} Q_{xx}^{-1} Q_{xz})\gamma}{2\sigma_1^2} \qquad (3.12)$$

$$\operatorname{plim}_{H_1} \hat{\lambda} = \operatorname{plim}_{H_1} [y' A_Z M_X y / y' A_Z M_X A_Z y] = 1 \qquad (3.22)$$

3.4.4. Asymptotic Size of a Test

According to Pesaran (1982, pg. 269), the size of a test is measured by the magnitude of its Type-I error which represents the probability of rejecting a true null hypothesis, i.e., the significance level. Test statistics such as the J- and Cox tests have correct and asymptotically equal sizes under the null hypothesis since both converge to the standard normal variate. However, the C-test of Davidson and MacKinnon (1981a) does not have a correct asymptotic size under the null hypothesis. As an illustration, their result is formally stated in the following theorem.

THEOREM 3.3.5

The C-test of H derived from the artificial regression model H_c: $y = X\hat{\beta}(1 - \lambda) + \lambda Z\hat{\gamma} + u$ do not have a correct asymptotic size even when H_0 is true.

PROOF

The proof of the above theorem is due to Davidson and MacKinnon 1981a, pg. 787 by considering the following consistent estimators of λ and the variance of λ based on least squares estimation of H_C.

$$\hat{\lambda} = \frac{y' A_z M_x y}{y' A_z M_x y + y' M_z A_x y} \tag{3.23}$$

$$\hat{V}_{\hat{\lambda}} = \hat{\sigma}^2_c [y' A_z M_x y + y' M_z A_x y]^{-1} \tag{3.24}$$

where $\hat{\sigma}^2_c$ represents the estimated variance of H_C. By assuming H_0 as the true model, the statistic $\sqrt{n}\,\hat{\lambda}$ can be written as:

$$\sqrt{n}\,\hat{\lambda} = [\sqrt{n}\ \beta' X' A_z M_x u_0 / \beta' X' M_z X \beta] \tag{3.25}$$

Since $u_0 \sim (0, \sigma_0^2 I_n)$ then $\sqrt{n}\,\hat{\lambda} \sim N(0, n V_{\hat{\lambda}})$

where

$$n V_{\hat{\lambda}} = n\ [\beta' X' A_z M_x A_z X \beta / (\beta' X' M_z X \beta)]^2. \tag{3.26}$$

On the other hand, the probability limit under H_0 of the estimated variance from regression H_C can be obtained from equation (3.24) as:

$$\text{plim}_{H_0} (n \hat{V}_{\hat{\lambda}}) = \sigma_0^2\ [\lim_{n\to\infty}(\beta' X' M_z X\beta / n)]^{-1} \tag{3.27}$$

By comparing the limits of (3.26) and (3.27) it can be established that the former is smaller than the latter, since

$$\beta' X' A_z M_x A_z X\beta \le \beta' X' M_x X\beta$$

Thus, the asymptotic variance of $\sqrt{n}\,\hat{\lambda}$ A under the true H_0 is smaller than its counterpart from the regression H_c which implies that asymptotically the value of Type-I error under the true H_0 will be smaller than the significance level for the C-test. In other words, the C-test does not have a correct asymptotic size.

3.4.5. A Small Sample Test: The JA-Test

Unlike the majority of specialized non-nested tests, the JA-test of Fisher and McAleer (1981) for H_0 is an exact test based on $\lambda = 0$ from the following compound model:

$$H_{JA}: y = X\beta(1-\lambda) + \lambda Z\hat{\gamma}_{10} + u_{JA},$$

where $Z\hat{\gamma}_{10} = (A_z\ A_x\ y)$ is the vector of predicted values obtained by regressing $\hat{y}_0$ on Z matrix, and $\hat{y}_0$ represents the vector of predicted values for y from H_0. Applying the same procedures, used earlier to derive the J-test, the JA-test may be defined as:

$$JA_0 = \frac{\sqrt{n}\ y'\, A_x\, A_z\, M_x\, y}{[\hat{\sigma}^2_{JA}\,(y'\, A_x\, A_z\, M_x\, A_z\, A_x\, y / n)]^{1/2}} \qquad (3.28)$$

where $\hat{\sigma}^2_{JA}$ denotes the estimated variance for the compound model H_{JA}.

Following the steps used in Theorem 3.3.3 for the J-test, it can be proved that asymptotically $JA_0 \sim N(0, 1)$ when H_0 is true. However, Godfrey (1981, 1983) and McAleer (1981, 1983) invoke the results of Milliken and Graybill (1970) to establish the small sample nature of JA_0 which is the least squares t-ratio for λ from H_{JA} above.

Milliken and Graybill show that if the models H_0 and H_1 contain only non-stochastic regressors, then this t-statistic will follow an exact Student's t-distribution with $(n - k_0 - 1)$ degrees of freedom under the true H_0, since the test variable, $Z\hat{\gamma}_{10}$, involves the product of non-stochastic variables and least squares estimators of the null hypothesis. (Here, k_0 represents the number of regressors in matrix X.)

On the other hand, Milliken and Graybill's result cannot be applied to derive an exact distribution for the J-test since its test variable, $Z\hat{\gamma}$, involves the least squares estimator, $\hat{\gamma}$, under H_1.

3.5. Consistency of Non-Nested Tests

Definition 11

A test statistic aimed at testing H_0 against H_1 is said to be consistent if it rejects a false H_0, in large samples, with unit probability.

If the test statistics such as J_0 or N_0 are asymptotically unbounded under H_1 then they are consistent statistics. Pereira (1977) provided the proof of consistency for a general form of Cox test while Pesaran (1982, pg. 271) proved the consistency of non-nested tests for linear models. Pesaran's approach will be followed here.

Theorem 3.4

The test statistics J_0 and N_{L_0}, defined respectively in (3.20) and (3.17), are consistent tests.

Proof

Asymptotic evaluation of the J-test in (3.20) under H_1 implies that

$$\text{plim}_{H_1} (J_0 / \sqrt{n}) = [\hat{\gamma}' (Q_{zz} - Q_{zx} Q_{xx}^{-1} Q_{xz}) \gamma /]^{1/2} > 0. \qquad (3.29)$$

Thus, J_0 is unbounded as $n \to \infty$ and certainly rejects H_0 when H_1 is true. Hence, the J-test is consistent.

Similarly, for the linearized Cox test, $\text{plim}_{H_1} (S_0 / \sqrt{n}) < 0$, see (3.21), and $(\hat{V}_0 / n) > 0$, since $(\hat{\beta}' X' M_z M_x M_z X \hat{\beta} n)$, $\text{plim}_{H_1} \hat{\sigma}_0^2$ and $\text{plim}_{H_1} \hat{\sigma}_{10}^2$ are positive definite.

Thus, $\text{plim}_{H_1} N_{L0} \to \infty$ and it leads to the definite rejection of a false null hypothesis. Hence, the linearized Cox test is a consistent test. Q.E.D.

3.6. Asymptotic Power of Non-Nested Tests

For nested hypotheses, the power of a test statistic refers to the probability of rejecting a false null hypothesis or simply it is the probability of not making a Type-II error. But for non-nested models, rejection of a false null does not characterize its power; rather, it involves both probabilities of rejecting the false and accepting the true models. Thus, in spite of the consistency of non-nested tests, an additional challenge for econometricians concerns the evaluation of the powers of various tests. Three distinct approaches discussed in the literature are outlined in the following subsections.

3.6.1. Local Alternative Approach

This approach involves the asymptotic evaluation of a test statistic under an alternative hypothesis, termed local alternative, which approaches the null hypothesis in an infinitely large sample. Such an approach examines the power of a test as its ability to distinguish between the null and alternative models for a given significance level as the sample size tends to infinity.

Pesaran (1982a) and Ericsson (1983) demonstrate the application of the local alternative approach to compare various non-nested tests which are consistent under a fixed alternative hypothesis. As a formal presentation of this approach, Pesaran's power statistics for the F- and J-tests are rederived below. (The notation here conforms with the present study and it may differ from the original work.)

To construct a local alternative, for H_0 defined in equation (3.2), Pesaran (1982) specifies the multicollinearity between the columns of X and Z matrices as:

$$Z = XB + n^{-1/2}\,\Delta + \phi \qquad (3.30)$$

where B and Δ are respectively k_0 xk_1 and nxk_1 non-zero matrices of constants and represents the terms of $o(n^{-1/2})$. To keep the asymptotic power of a test greater than its value of Type-I error, it is assumed that $\lim_{n\to\infty}[(\Delta' M_x \Delta)/n] = Ɖ$, here, Ɖ is a non-zero, finite matrix of order k_1 xk_1.

Substituting the value of Z from (3.30) in H_1, defined in equation (3.3), a local alternative denoted as H_L, may be written as:

$$H_L: y = XB\,\gamma + n^{-1/2}\,\Delta\gamma + \phi\,\gamma + u_1 u. \qquad (3.31)$$

Without loss of generality, it may be assumed that all the columns of Z and X are non-overlapping; otherwise, Z, Δ, and γ can be partitioned accordingly and H_L will be defined in terms of the non-overlapping variables only. Note that in the above specification $H_1 \to H_0$ as $n \to \infty$ and it becomes increasingly difficult for a test statistic to distinguish between H_0 and H_L for an increasing sample size. Thus, the asymptotic powers of the F- and J-tests under such a local alternative are obtained in the following subsections.

3.6.2. Asymptotic Power of the F-Test

The F-test defined in equation (2.5) can be rewritten as:

$$F_0 = y' M_x Z(Z' M_x Z)^{-1} Z' M_x y / \hat{\sigma}_0^2 \qquad (3.32)$$

Let $\varsigma = (Z' M_x Z)^{-1/2} Z'M_x\, y / \sigma_1$. Then, under H_L, after substitution of y and Z from the above equations and retaining terms of only higher order in probability, the expression for ς reduces to the following expression.

$$\varsigma = (\Delta' M_x \Delta / n)^{-1/2} [\, (\Delta' M_x \Delta \gamma / n) + (\Delta' M_x u_1 / \sqrt{n}\,) / \sigma_1 \qquad (3.33)$$

Using the assumption $u_1 u \sim N(0, \sigma_1^2 I_n)$, it is established that ς is asymptotically distributed as $N(\mu, I_{K_1})$ under H_L where $= (\Delta' M_x \Delta / n)^{1/2} \gamma / \sigma_1$ and I_{K_1} is K_1 dimensional identity matrix.

Note that $F_0 = \varsigma'\varsigma / (\hat{\sigma}_0^2 / \sigma_1^2)$ and $\text{plim}_{H_L} (\hat{\sigma}_0^2 / \sigma_1^2) = 1$ implies that F_0 and $\varsigma'\varsigma$ are asymptotically equivalent under H_L. The normality of ς with mean μ establishes that $\varsigma'\varsigma$ or F_0 follows a non-central χ^2 with k_1 degrees of freedom and with asymptotic non-centrality parameter, denoted as η_F, given by:

$$\eta_F = \lim_{n\to\infty} \mu'\mu = \gamma' \, \text{Đ} \gamma / \sigma_1^2 \qquad (3.34)$$

Pesaran (1982a, pg. 1294) defines the power function of the F-test as:

$$\begin{aligned} P_F &= \lim_{n\to\infty} P[F_0 \geq X_\alpha^2 (k_1, 0) \mid H_L] \\ &= P\, [\chi^2 (k_1, \eta_F) \geq (k_1, 0)] \end{aligned} \qquad (3.35)$$

Where α in (3.35) denotes the significance level. P_F clearly depends on the values of its degrees of freedom, k_1, and non-centrality parameter, η_F. This power function can be computed given the degrees of freedom.

3.6.3. Asymptotic Power of the J-Test

The J-test for H_0 is defined in equation (3.20) as:

$$J_0 = y' A_z M_x y / [\hat{\sigma}_J^2 (y' A_z M_x A_z y / n)]^{1/2}$$

Substituting y and Z values under H_L, in the numerator of J_0 and eliminating the terms of lower order in probability results in the following expression:

$$y' A_z M_x y = (\gamma'\Delta' M_x \Delta \gamma / n) + (\gamma' \Delta' M_x u_l / \sqrt{n}) \tag{3.36}$$

From the normality assumption of u_l, it follows that under H_L the expression in (3.36) is asymptotically distributed as follows:

$$y' A_z M_x y \sim N(\gamma' Đ \gamma, \sigma_1^2 \gamma' Đ \gamma)$$

On the other hand, plim H_L $[\hat{\sigma}_J^2 (y' A_z M_x A_z y / n)]^{1/2} = \sigma_1 (\gamma' Đ \gamma)^{1/2}$. Thus, under H_L, the asymptotic distribution of J_0 may be written as:

$$J_0 \sim N(\gamma' Đ \gamma, / \sigma_1^2]^{1/2}, 1),$$

which implies that J_o^2 follows a non-central X^2_1 distribution under H_L, and the non-centrality parameter, η_J, equals η_F. Thus, the asymptotic power of the J-test is analogously defined as:

$$\begin{aligned} P_J &= \lim_{n\to\infty} P[J_0 \geq X^2_\alpha (1, 0) \mid H_L] \\ &= P[X^2 (1, \eta_J) \geq X^2_\alpha (1, 0)] \end{aligned} \tag{3.37}$$

THEOREM 3.5.1

Under the local alternative, H_L, the J-test has higher asymptotic power than the F-test when H_1 has more than one distinct regressors, i.e., $P_J > P_F$ if $k_1 > 1$.

PROOF *(Pesaran 1982a, pg. 1297-1298)*

Das Gupta and Perlman (1974) show that a non-central X^2 statistic has higher power with lower degrees of freedom. Thus, for $\eta_F = \eta_J$, $P_F > P_J$, if $k_1 > 1$ and $P_F = P_J$, if $k_1 = 1$.

Pesaran (1982a, pg. 1294-96), similarly, derives the asymptotic power for the Cox-test which also equals the asymptotic power of the J-test. Thus, asymptotically Cox and J-tests are equivalent.

3.6.4. Fixed Alternative Approach

To evaluate the power of a test statistic Davidson and MacKinnon (1982, pg. 556) propose "...keep[ing] the alternative H_1 fixed, and merely let the process generating y approach H_0." They argue that, unlike the local alternative approach, their approach will be useful when H_0 and H_1 are both false but the true data generating process is in the neighborhood of H_0.

Let H_T, the true data generating process, be specified as:

$$H_T: y = X\beta_0 + \phi + u, \tag{3.38}$$

where $X\beta + \phi \neq X\beta_0$ for a fixed parameter β_0, ϕ is a non-stochastic vector of order $n^{-1/2}$ and $\phi'\phi$ tends to a finite limit as $n \to \infty$ and $u \sim N(0, \sigma^2 I_n)$. The power of a test statistic, in this case, is represented by its limiting value under H_T. Davidson and MacKinnon (1982) consider the evaluation of the P- and Cox tests for non-linear models and prove the asymptotic equivalence of both tests under this approach.

Using their techniques, the fixed alternative approach is applied in the following to evaluate the power of F- and J- tests for linear models H_0 and H_1. Unlike the earlier approach, the comparison here proves to be inconclusive.

Fixed Alternative Approach and the Power of the F-Test

From (3.32) the F-test is:

$$F_0 = \varsigma'_0 \varsigma_0 / (\hat{\sigma}_0^2 / \sigma^2),$$

where $\varsigma_0 = (Z' M_x Z)^{-1/2} Z' M_x y / \sigma$. Substituting y from H_T and using the normality assumption regarding u, it can be asymptotically established that

$$\varsigma_0 \sim N[(Z' M_x Z)^{-1/2} Z' M_x \phi / \sigma, I_{K_1}]$$

Also note that $\text{plim}_{H_0}\ \hat{\sigma}_0^2 = \sigma^2$. Thus, F_0 and $\varsigma'_0\ \varsigma_0$ asymptotically equivalent under H_T and have non-central x^2 distribution with k_1 degrees of freedom and the asymptotic value of non-centrality parameter is:

$$\eta_F = \lim_{n\to\infty} \phi'\ , M_x Z(Z' M_x Z)^{-1} Z' M_x\ \phi / \sigma^2$$

Note that such a limit for η_F exists by the definition of ϕ see Davidson and MacKinnon (1982, pg. 557). Thus, the power of the F-test, P_F, may be defined in exactly the same way as above.

Fixed Alternative Approach and the Power of the J-Test

Consider the following expression of the J-test:

$$J_0 = \sqrt{n}\ y' A_z M_x y /\ [\hat{\sigma}_J^2\ (y' A_z M_x A_z y)]^{1/2}$$

Again, substituting y from H_T and using the normality of u implies the following asymptotic distribution of J_0 under H_T: $J_0 \sim N(\beta_0' X'\beta\ A_z M_x \phi / [\sigma^2 (\beta_0'\beta X' A_z M_0 A_z X\beta_0)]^{1/2}, 1)$ and J_0 is, therefore, an asymptotic non-central X_1^2 with limiting non-centrality parameter given by:

$$\eta_J = \lim_{n\to\infty} [\ \phi' M_x A_z X\beta_0\ \beta_0' X' A_z M_x \phi / \sigma^2 (\beta_0' X' A_z M_x A_z X \beta_0)]$$

Again the limit for η_J, will exist which ensures the existence of the power, P_J, for the J-test. In general, the direct comparison of P_J and P_F in this case is inconclusive since the degrees of freedom as well as non-centrality parameters of the asymptotic X^2 distributions for both tests are different.

Davidson and MacKinnon (1982), however, show that the Cox and P-tests have the same mean and unit variance under H_T for non-linear models and are asymptotically equivalent. Similarly, for linear models, the Cox and J tests have the same asymptotic power when both H_0 and H_1 are false but $H_T \to H_0$ as $n \to \infty$.

3.6.5. Bahadur's Approach

The third approach due to Bahadur (1960, 1967) which has been used by Pesaran (1984) to evaluate the asymptotic power of non-nested tests. Pesaran defines Bahadur's approach, in contrast to the local alternative approach. In the local approach the probability of type I error is held fixed but the alternative hypothesis is allowed to approach the null. Bahadur's approach holds the alternative hypothesis fixed, but allows the type I error to tend to zero as the sample size increases. ...a test is asymptotically efficient to another if its approximate slope is greater. (Pesaran 1984, pg. 247) The approximate slope of a test statistic, say N_0, and under null hypothesis its plim(N_0^2/ n). Thus, from (3.29) the approximate slope of J-test may be easily given by:

$$\text{plim}_{H_1} (J_0^2 / n) = [\lim_{n\to\infty}(\gamma' Z' M_x Z\gamma / n) / \sigma_1^2], \tag{3.39}$$

and the approximate slope of linearized Cox test may be defined as:

$$\text{plim}_{H_1} (N_{L_0}{}^2 / n) =$$

$$[(\lim_{n\to\infty} \gamma' Z'(M_x + A_x M_z A_x) Z\gamma / n)^2 / 4 \lim_{n\to\infty}(\hat{V}_0 / n)]. \tag{3.40}$$

Unfortunately, the comparison of (3.39) and (3.40) is not conclusive in terms of relative efficiency. However, Pesaran (1984, pg. 250-51) proves the relative efficiency of the Cox test, for a specific linear form, over the J and F-tests which have same approximate slope as obtained in (3.39).

Finally, note that none of these approaches prove to be superior than others after ranking various tests. However, Pesaran (1986) suggests the use of each approach under different circumstances. He suggests that Bahadur's approach is only appropriate for testing globally non-nested models and may not be generally useful to compare tests of linear regression models since "linear regression models are either nested or partially non-nested". On the other hand, the local alternative approach is suitable for such situations. However, if it is not possible to define a "specific alternative", then the fixed alternative approach should be followed to evaluate the asymptotic powers of non-nested tests.

3.7. Conclusion

This chapter describes the importance of asymptotic theory in the context of non-nested hypothesis testing and errors-in-variables models. The basic concepts of stochastic limits and orders in probability are first defined and then applied to non-nested hypothesis tests. Such applications and proofs are not necessarily original in content, but, rather, provide a technical perspective for the present study. The proofs and results regarding the asymptotic distributions of various tests, the equivalence of the J- and linearized Cox test, under H_0, and incorrect asymptotic size of the C-test follow from Davidson and MacKinnon (1981a). The proofs for consistency of the J- and Cox test are reproduced from Pesaran (1982). Three alternative approaches to evaluate asymptotic powers of the F-, J-, and Cox tests are used. The local alternative approach of Pesaran (1982a) proves the asymptotic efficiency of the J- and Cox tests over the F-test when the number of non-overlapping variables exceeds one. Alternatively, Davidson and MacKinnon's (1982) fixed alternative approach demonstrates the equality of powers for the Cox and P-tests. Their approach has been followed here to demonstrate that no absolute ranking of the powers of the F- and J-tests exist for linear regression models. Finally, Bahadur's approximate slope criterion used by Pesaran (1984) proves the relative efficiency of the Cox test over the J- and F- tests. However, his result is limited to the specific linear approximation chosen for the Cox test and, thus, cannot be generalized.

In general, it is not possible to declare a single test statistic as the most powerful test among the class of existing non-nested tests even asymptotically. Moreover, no concrete evidence exists towards the relative performance of IV-based tests suggested by Godfrey (1983), Ericsson (1983), MacKinnon et al. (1983), and Dastoor and McAleer (1985). However, the importance of such statistics for testing EVM in the non-nested framework cannot be overlooked. Thus, the focus of the next chapter is to evaluate the performance of separate hypothesis tests for errors-in-variables regression models.

The existing IV-based non-nested tests will be applied to test such specifications. An explicit effort will be made to evaluate and compare asymptotic powers of various test statistics. Following Pesaran's (1986) observation, the local and fixed alternative approaches will be used for these power comparisons.

Chapter 4

NON-NESTED TESTS AND EVM – MODELS

4.1. INTRODUCTION

It has been emphasized earlier that two major econometric issues, non-nested tests and errors-in-variables models, may occur together in applied economics. The existing econometric literature, however, lacks evidence on the behavior of non-nested tests when one or more competing regression models include mismeasured regressors. This chapter, therefore, provides the asymptotic analysis of existing non-nested procedures to test one or more errors-in-variables models. Some modified tests are suggested to obtain consistent and asymptotically efficient non-nested tests for such situations.

This chapter is organized in the following six sections. The next section presents an econometric specification of the errors-in-variables models and the second section investigates the behavior of various non-nested tests when the null hypothesis is mismeasured. The third section presents, IV estimators for the mismeasured null hypothesis and uses IV estimators to modify some of the non-nested tests and evaluate their asymptotic size and power behavior. The fourth section presents modified tests when null and alternative models are both mismeasured and IV estimators are employed. The fifth section discusses the choice of an alternative in relation to IV matrix and the last section concludes this chapter by summarizing the major results.

4.2. ERRORS-IN-VARIABLES MODELS

Economists typically analyze economic relationships subject to measurement errors within the classical assumptions such as linearity of the slope parameters and Gaussian "white noise" error terms. This classical model with the full ideal conditions, restated here from section 1.1, is given by:

$$y = X\beta + u_0$$

$$\lim_{n\to\infty} \frac{X'X}{n} \text{ exist and is non-singular.}$$

$E(u_0) = 0$, where E denotes the expectation operator.

$u_0 \sim N(0, \sigma_0^2 I_n)$.

Here, y is an n-valued vector of observations on the dependent variable; X, an nxk_0 matrix of non-stochastic regressors of full column rank; u_0, an n-valued vector of stochastic error terms; β and σ_0^2, unknown position and shape parameters, respectively, with n representing the number of sample observations.

It is well known that the non-stochastic nature of the design matrix, X, in the classical model is fundamental to the consistency property of the least squares estimator of β. In particular, its non-randomness precludes any small- or large-sample correlation with the error term, an assumption that may well be violated when one or more of the regressors is mismeasured or unobservable.

Alternatively, the errors-in-variables model specifies the exact form of mismeasurement within the following framework.

$$y = X_T\beta X_T\beta + u_T \tag{4.1}$$

$$X = X_T + V_0 \tag{4.2}$$

$$y = X\beta + u_0 \tag{4.3}$$

$$u_0 = u_T - V_0\beta. \tag{4.4}$$

Equation (4.1) represents the "true", albeit unobservable regression model, whereas (4.3) relates the model in terms of the observable, random values for X. Here, u_T is an nxl vector of the structural, stochastic disturbance terms in which $u_T \sim N(0, \sigma_T^2 I_n)$, V_0 is an nxk_0 matrix of mismeasurement errors on X, and X_T is an nxk_0 matrix of "true", non-stochastic, unobservable regressors. Note that the error in measuring the X matrix is presumed to enter additively. The EVM will also include additional assumptions such that $E(V_0) = 0$, an nxk_0 matrix of zeroes and $E(u_T'v_i) = 0$, for each v_i column vector in V_0.

These latter assumptions imply that there is no persistent under- or over-mismeasurement to X and that mismeasurement and structural error terms are uncorrelated. Furthermore, note that, depending, on the number of mismeasured regressors, the rank of V_0 may be less than k_0, in that V_0 may include columns of zeroes.

In practice, the estimators of β and σ_0^2 can only be generated from the observed values in (4.3), although (4.1) represents the DGP. The inconsistency of the ordinary least squares estimators of (4.3) is well known (see section 3.2), and instrumental variables estimation is one proposed alternative. The next section begins the discussion of non-nested testing of mismeasured regression models by showing that existing separate tests based on the least squares method will produce misleading or incorrect inferences when one of the models is subject to mismeasurement error.

4.3. Properties of Non-Nested Tests in Mismeasured Regression

Attention will be focused here on the consequences, if any, of ignoring mismeasurement problems and proceeding with the usual separate hypothesis tests by evaluating possible distortions to the sampling distributions of various non-nested tests such as the Cox, J-, JA-, and encompassing principle approaches when the regressors in any one model are mismeasured or unobservable. Mismeasured regressands are discussed only in passing, since this situation will not affect the fundamental results of this section.

Consider the following separate, linear regression models:

$$y = X_T\beta + u_T$$
$$y = Z\gamma + u_1,$$

where X_T is unobservable, although proxy observations exist for X_T that are presumed to behave according to the EVM model in (4.1) through (4.4). The problem, in terms of the observed data for estimation and testing, is to discern between the two models:

$$H_0: y = X\beta + u_0 \tag{4.5}$$
$$H_1: y = Z\gamma + u_1 \tag{4.6}$$

where y is nxl vector of observations on the dependent variable; X and Z are nxk_0 and nxk_1 observed matrices of stochastic and fixed regressors, respectively, and both have full column rank; u_0 and u_1 are nxl unobserved random vectors such that $u_0 \sim N(0, \sigma_0^2 I_n)$ and $u_1 \sim N(0, \sigma_1^2 I_n)$; β and γ are, respectively, k_0 xl and k_1 xl unknown parameter vectors while σ^2_0 and σ^2_1 are unknown scalars.

To formalize the non-nested structure of H_0 and H_1, consider the partitioning of matrices X and Z such that $X = [X_1 \mid \overline{X}]$ and $Z = [X_1 \mid \overline{Z}]$, where X_1 is an nxr matrix ($r < k_0$ and $r < k_1$) of common variables in both models, while and are, respectively, $(k_0 - r)$ and $(k_1 - r)$ columns of distinct regressors. If r = 0, then both models are globally non-nested; if $0 < r < k_0, k_1$ then the models are partially non-nested; if $r = k_0$ or $r = k_1$ then the models are nested. The role of r is of special importance in the forthcoming analysis. In addition, the following technical requirements complete the specification of these models.

1) To obtain the normality of estimators and hypothesis tests for H_0 and H_1, it is customary to assume that u_0 and u_1 are normally distributed i.e.. $u_0 \sim N(0, \sigma_0^2 I_n)$ and $u_1 \sim N(0, \sigma_1^2 I_n)$. However, the normality of u_0 will depend on the requirements of normality for u_T and v_i (see equation 4.4), i.e., $u_T \sim N(0, \sigma_T^2 I_n)$ and $v_i \sim N(0, \sigma_{v_0}^2 I_n)$.

2) To ensure the existence of estimators for the unknown parameters of both models, following assumptions are the part of classical regression modles: plim $\dfrac{X'X}{n} = Q_{xx}$, a finite, positive definite, non-singular matrix, and $\lim_{n\to\infty} \dfrac{Z'Z}{n} \underset{n \to \infty}{\lim} = Q_{zz}$, also a finite positive definite, non-singular matrix. For non-nested models, the regressors in X and Z are assumed to be non-orthogonal: plim $\dfrac{X'Z}{n} = Q_{xz}$, $Q'_{xz} = Q_{zx}$ exists as a non-zero matrix.

3) To further specify the relationships between structural and random disturbances as well as between the regressors and different error terms, the following assumptions and results will be invoked.

(i) $\text{plim}\,\frac{u'_0 u_0}{n} = \sigma_0^2$, where $\sigma_0^2 = \sigma_T^2 + \beta' Q_{V_0} \beta$, assuming that

$\text{plim}\,\frac{V'_0 V_0}{n} = Q_{V_0}$ is a finite matrix of order $k_0 x k_0$, and

$\text{plim}\,\frac{V'_0 u_T}{n} = 0$, a column vector of zeroes.

(ii) $\text{plim}\,\frac{u'_1 u_1}{n} = \sigma_1^2$ which follows directly from the classical structure of H_1.

(iii) $\text{plim}\,\frac{X' u_0}{n} = -\, Q_{V_0} \beta$, assuming $\text{plim}\,\frac{X'_T V_0}{n} = 0$, a null matrix of order $k_0 x k_0$, and $\text{plim}\,\frac{X'_T u_T}{n} = 0$, a $k_0 x 1$ vector of zeroes.

(iv) $\text{plim}\,\frac{Z' u_0}{n} = \text{plim}\,\frac{Z' u_1}{n} = 0$, since Z is a matrix of fixed regressors.

The linearized Cox test, J-test, JA-test, and Wald Encompassing test (WET), whose values are reproduced below, will be considered in the following analysis. The linearized Cox test of H is given by:

$$S_0 = -\,(\sqrt{n}/2)\,[\,(\hat{\sigma}_{10}^2 - \hat{\sigma}_1^2)\hat{\sigma}_1^2\,] \tag{4.7}$$

The J-test is given by:

$$\sqrt{n}\,\hat{\lambda}_{J_0} = \sqrt{n}\,[\,\hat{\gamma}' Z' M_x y \,/\, \hat{\gamma}' Z' M_x Z \hat{\gamma}\,] \tag{4.8}$$

The JA-test of H_0 is based on the following test statistic:

$$\sqrt{n}\,\hat{\lambda}_{JA_0} = \sqrt{n}\,[\hat{\gamma}'_{10} Z' M_x y \,/\, \hat{\gamma}'_{10} Z' M_x Z \hat{\gamma}_{10}]. \tag{4.9}$$

The WET statistic by Mizon (1984), the numerical analogue of the F-test or the Wald statistic for complete parametric encompassing by Mizon and Richard (1986, pg. 669), may be written as follows:

$$\hat{\eta}_{W_0} = \frac{\varsigma'_0 . \varsigma_0}{\hat{\sigma}_0^2 / \sigma_0^2} \tag{4.10}$$

Here in (4.10), $\varsigma_0 = \dfrac{(\overline{Z}' M_x \overline{Z})^{-1/2} \overline{Z}' M_x y}{\sigma_0}$

$$\hat{\beta} = (X'X)^{-1} X'y$$

$$\hat{\gamma} = (Z'Z)^{-1} Z'y$$

$$\hat{\gamma}_{10} = (Z'Z)^{-1} Z'\hat{y}_0$$

$$\hat{\sigma}_0^2 = \frac{y' M_x y}{n}$$

$$\hat{\sigma}_1^2 = \frac{y' M_z y}{n}$$

$$\hat{\sigma}_{10}^2 = + \frac{1}{n} \hat{\beta}' X' M_z X \hat{\beta}$$

$$\hat{y}_0 = A_x y$$

$$M_x = I_n - A_x$$

$$M_Z = I_n - A_z$$

$$A_x = X(X'X)^{-1} X'$$

$$A_z = Z(Z'Z)^{-1} Z'.$$

The statistics $\sqrt{n}\, \hat{\lambda}_{J_0}$ and S_0 were shown earlier to have following statistical properties when the non-nested models H_0 and H_1 satisfy the ideal Gaussian conditions.

(i) Under H_0, S_0 and $\sqrt{n}\, \hat{\lambda}_{J_0}$ asymptotically converge to zero.

(ii) The statistics are asymptotically equivalent under H_0.

(iii) Both statistics are consistent.

The asymptotic convergence and equivalence of these two tests under the mismeasured H_0 of (4.6) and consistency properties for only these two test statistics are studied in the following theorem. However, these results will easily extend to other non-nested tests.

THEOREM 4.1

When H_0 is an EVM, the statistics S_0 and $\sqrt{n}\ \hat{\lambda}_{J_0}$ are neither asymptotically zero mean nor asymptotically equivalent. However, both statistics are consistent.

PROOF

LINEARIZED COX TEST

Examination of the linearized Cox test in (4.7) indicates that this statistic will not converge to zero if $\text{plim}_{H_0}\ \hat{\sigma}_1^2 \neq \text{plim}_{H_0}\ \hat{\sigma}_{10}^2$. When H_0 is the EVM, then

$$\text{plim}_{H_0}\ \hat{\sigma}_1^2 = \hat{\sigma}_0^2 + \beta'\,(\,Q_{XX} - Q_{XZ}\,Q_{ZZ}^{\ -1}Q_{ZX}\,)\beta - 2\,\beta' Q_{V_0}\,\beta \neq 0 \quad (4.11)$$

$$\text{plim}_{H_0}\ \hat{\sigma}_{10}^2 = \sigma_0^2 - \beta'\ Q_{V_0}\,Q_{XX}^{\ -1}\ Q_{V_0}\,\beta + \beta'\,[\,I_{K_0} - Q_{V_0}\,Q_{XX}^{\ -1}]$$
$$[Q_{XX} - Q_{XZ}\,Q_{ZZ}^{\ -1}Q_{ZX}]\,[\,I_{K_0} - Q_{XX}^{\ -1}Q_{V_0}\ ]\beta \neq 0 \quad (4.12)$$

Thus, $\text{plim}_{H_0}\ \hat{\sigma}_1^2 \neq \text{plim}_{H_0}\ \hat{\sigma}_{10}^2$ *proves that* $\text{plim}_{H_0}\ S_0 \neq 0$

J-TEST

To obtain $\text{plim}_{H_0}\ (\sqrt{n}\ \hat{\lambda}_{J_0})$, consider the following equations:

$$\text{plim}_{H_0}\ \hat{\gamma} = Q_{ZZ}^{\ -1}Q_{ZX}\,\beta \neq 0. \quad (4.13)$$

$$\text{plim}_{H_0}\ [\,Z' M_X\,y\,/\,n\,] = Q_{ZX}\,Q_{XX}^{\ -1}\ Q_{V_0}\,\beta \quad (4.14)$$

$$\text{plim}_{H_0}\ [Z'\,M_X\,Z\,/\,n] = Q_{ZZ} - Q_{ZX}\,Q_{XX}^{\ -1}\,Q_{XZ} \neq 0 \quad (4.15)$$

Using (4.13) through (4.15), it can be established that plim_{H_0} $(\hat{\gamma}'Z'M_x y / n)$ and plim_{H_0} $(\hat{\gamma}'Z'M_x Z\hat{\gamma} / n)$ are finite and non-zero. Thus,

$$\text{plim}_{H_0} (\sqrt{n}\,\hat{\lambda}_{J_0}) \neq 0.$$

Second, $\text{plim}_{H_0} S_0 \neq \text{plim}_{H_0} \sqrt{n}\,\hat{\lambda}_{J_0}$, since $[-\sqrt{n}/2\,(\hat{\sigma}_{10}^2 - \hat{\sigma}_1^2)]$ exceeds $\text{plim}_{H_0} [-\sqrt{n}\,(\hat{\gamma}'Z'M_x y / n)]$ by a positive quantity:

$$\frac{1}{2}\beta' Q_{V_0} Q_{xx}^{-1} Q_{xz} Q_{zz}^{-1} Q_{zx} Q_{xx}^{-1} Q_{V_0}, \tag{4.16}$$

and $\text{plim}_{H_0} \hat{\sigma}_1^2$ and $\text{plim}_{H_0} (\hat{\gamma}'Z'M_x Z\hat{\gamma} / n)$ are non-zero, finite quantities. Thus, under H_0, S_0 and $\sqrt{n}\,\hat{\lambda}_{J_0}$ are not asymptotically equivalent.

Finally, the consistency of these tests directly follows from the Theorem 3.4, since H_1 is exactly the same model in both cases - the correctly measured hypothesis. Similarly, asymptotically non-zero mean for the JA-test may be obtained in the following.

JA-TEST

Using (4.14) and (4.15) and the following result, it can be seen that $\text{plim}_{H_0} (\sqrt{n}\,\hat{\lambda}_{JA\,0}) \neq 0$:

$$\text{plim}_{H_0} \hat{\gamma}_{10} = Q_{zz}^{-1} Q_{zx} [\, I_{k_0} - Q_{xx}^{-1} Q_{V_0}]\, \beta \neq 0. \tag{4.17}$$

From this result, it may also be noted that Godfrey's (1983) T-test based on least squares estimators - a negative valued, numerical analogue of the JA-test - will converge to a non-zero limit under a mismeasured H_0.

From (4.14) and (4.15) it can be readily seen that $\text{plim}_{H_0} \zeta_0 \neq 0$. Thus, the WET statistic, $\hat{\eta}_{W_0}$, follows an asymptotically non-central $\chi^2_{K_1}$-r when H_0 is an EVM and the same will be true for other non-nested testing procedures such as the conventional F-test, R-test of Dastoor (1983), and variance encompassing test of Mizon (1984).

Analogously, if H_0 and H_1 are both specified as EVM then the statistics $T_1 / \sqrt{n}$, $\sqrt{n}\ \hat{\lambda}_{J_1}$, $\sqrt{n}\ \hat{\lambda}_{JA1}$, and $\hat{\eta}_{W_1}$ will also converge to non-central limits under H_1. Furthermore, these results may be extended to the multiple analogues of these test statistics.

The results of this section are summarized here. Recognizing that the two models — one, a correctly measured regression model and the other, an EVM — will alternately play the role of the tested hypothesis (the null model) and the alternative hypothesis, and that only one of them may be "true", the least-squares-based non-nested tests have the following properties. If the EVM is the null hypothesis when in realistic it is "false", it will be rejected asymptotically as being true by a correctly-measured alternative hypothesis. A correctly-measured null model which is the true model will fail to be rejected by an EVM alternative. However, the "truth" of an EVM serving as H_0 will never be correctly ascertained asymptotically, a result that can be traced directly to inconsistency of the least-squares estimators of β in errors-in-variables models. Moreover, it is either undesirable or impractical to impose mean corrections to the Cox and linear embedding procedures (or centrality corrections to the WET test, for example) since the appropriate terms for the corrections are functions of unobservable parameters. Finally, the linearized Cox, J-, JA-, and WET statistics for H_0 are neither asymptotically equivalent nor can be ranked in terms of their asymptotic means under the mismeasured H_0.

Next the case of a mismeasured regresses and is considered; it is shown that the asymptotic properties of the non-nested statistics will be unaffected due to such distortions. Let the observed vector y be mismeasured, such that

$$y = y_T + e, \tag{4.18}$$

where $e \sim N(0, \sigma_e^2 I_n)$. The non-nested regression models H_0 and H_1 can now be written as:

$$H_0\!: y = X\beta + u_0 - e \tag{4.19}$$
$$H_1\!: y = Z\gamma + u_1 - e \tag{4.20}$$

The following assumptions are typically made about EVM's of this sort:

plim $(e'u_0 / n)$ = plim $(e'u_1 / n) = 0$, plim $(X'u_0 / n)$ = plim $(X'e / n)$ = plim $(X'u_1 / n) = 0$, a null vector of order k_0 , and plim $(Z'u_0 / n)$ = plim $(Z'e / n)$ = plim $(Z'u_1 / n) = 0$, a null vector of order k_1. These assumptions provide zero pairwise asymptotic correlation between the regressors (presumed here to be non-stochastic) and each error vector, and between each pair of error terms in H_0 and H_1. The remaining classical regression model assumptions presented earlier, as well as the conditions applying to the non-orthogonality of X and Z, will be used in the following, but, for brevity, are not reproduced here.

Now, unlike the case of mismeasured regressors, it is shown below that the linearized Cox, J-, and JA-tests asymptotically converge to zero under H_0, in equation (4.19).

THEOREM 4.2

For non-nested EVM characterized with non-stochastic regressors but mismeasured dependent variables, the Cox, J-, and JA-tests are asymptotically zero under H_0.

PROOF LINEARIZED COX TEST

$$\text{plim}_{H_0}\ S_0 = -\frac{\sqrt{n}}{2}\ \text{plim}_{H_0}\ \frac{\hat{\sigma}^2_{10}\ \hat{\sigma}^2_1}{\hat{\sigma}^2_1}$$

$$= -\frac{\sqrt{n}}{2}\ \text{plim}_{H_0}\ \frac{[y'(A_z - A_x A_z A_x)\, y / n\,]}{(y' M_z\, y / n)}$$

$$= -\frac{\sqrt{n}}{2}\ \text{plim}\ \frac{[(X\beta + u_0 - e)'\,[A_z - A_x A_z A_x]\,(X\beta + u_0 - e) / n\,]}{[(X\beta + u_0 - e)'\, M_z\ (X\beta + u_0 - e) / n]}$$

$$= -\frac{\sqrt{n}}{2}\ \frac{0}{\beta'\,(Q_{xx} - Q_{xz} Q_{zz}^{-1} Q_{zx})\beta}$$

$$= 0$$

J-TEST

$$\text{plim}_{H_0}\left(\sqrt{n}\,\hat{\lambda}_{J_0}\right) = \sqrt{n}\ \text{plim}_{H_0} \frac{[y' - A_z M_x y / n]}{[y' - A_z M_x A_z y / n]}$$

$$= \frac{\sqrt{n}\ .0}{\beta'\,(Q_{xz}\,Q_{zz}^{-1}Q_{zx} - Q_{xz}\,Q_{zz}^{-1}Q_{zx}\,Q_{xx}^{-1}Q_{xz}\,Q_{zz}^{-1}Q_{zx})\beta}$$

JA-TEST

$$\text{plim}_{H_0}\ \sqrt{n}\,\hat{\lambda}_{JA_0} = \sqrt{n}\ \text{plim}_{H_0} \frac{[y'\ A_x A_z M_x y / n]}{[y'\ A_x A_z M_x A_z A_x y / n]}$$

$$= \frac{\sqrt{n}\ .0}{\beta'\,(Q_{xz}\,Q_{zz}^{-1}Q_{zx}\ {}^{-}Q_{xz}\,Q_{zz}^{-1}Q_{zx}\,Q_{xx}^{-1}Q_{xz}\,Q_{zz}^{-1}Q_{zx})\beta}$$

$$= 0$$

COROLLARY 4.2

The WET statistic is asymptotically central $\chi^2_{K_1}$-r when H_0 is characterized by a mismeasured regress and non-stochastic regressors.

PROOF

Let $\zeta_0 = [(\overline{Z}'M_x\overline{Z})^{-1/2}\,\overline{Z}'M_x y / (\sigma_0^2 + \sigma_e^2)^{1/2}]$. Then, substituting Xβ + u_0 - e for y and using the normality assumptions regarding u_0 and e, it can be asymptotically established that $\zeta_0 \sim N(0,\ I_{k_1} - r)$. Thus, $\zeta_0\,'\,\zeta_0$ is asymptotically – r.

Also note that $\text{plim}_{H_0}\ \hat{\sigma}_0^2 = \sigma_0^2\ \ \sigma_e^2$. Thus, $\zeta_0\,'\,\zeta_0$ and $\hat{\eta}_{W_0}$ are asymptotically equivalent. Hence, $\hat{\eta}_{W_0}$ is asymptotically $\chi^2_{k_1}$ - r under H_0.

These results are of considerable interest to applied econometricians, since the non-nested tests will fail to reject a true model when only the regress and in non-nested regression models is mismeasured. In fact, it can be shown that these tests are also consistent, a proof that is a straight forward extension of the steps outlined in chapter 3. In the following theorem, the J-test is used to demonstrate that the separate hypothesis tests are of the correct size asymptotically. In terms of power, the Cox, J-, JA-tests, based on the local alternative approach, can be shown to be asymptotically equivalent and superior to the WET or F-test, as in the case of correctly observed regression models (again, see chapter 3). The following theorem, then, allows applied researchers to confidently proceed with separate hypothesis tests when the only source of mismeasurement occurs in the dependent variable.

THEOREM 4.3

For non-nested models characterized with non-stochastic regressors but mismeasured dependent variables, the J-test is of the correct size asymptotically.

PROOF

Recall that the J-test is derived from the compound model:

$$H_J: y = X\beta\,(1 - \lambda) + \lambda\, Z\,\hat{\gamma} + u_J$$

As usual, $\hat{\lambda}_{J_0} = (y'\, A_z\, M_x\, y \,/\, y'\, A_z\, M_x\, A_z\, y)$. Under H_0, can be written as:

$$\sqrt{n}\,\hat{\lambda}_{J_0} = \frac{(X\beta + u_0 - e)'A_z\, M_x\,(u_0 - e)\,/\,\sqrt{n}}{(X\beta + u_0 - e)'A_z\, M_x\, A_z\,(X\beta + u_0 - e)\,/\,n}$$

Dropping terms of lower order in probability gives

$$\sqrt{n}\,\hat{\lambda}_{J_0} = \frac{\beta'X'\, A_z\, M_x(u_0 - e)\,/\,\sqrt{n}}{\beta'X'\, A_z\, M_x\, A_z\, X\beta\,/\,n}$$

Defining $\sigma_y^2 = \sigma_0^2 + \sigma_e^2$ the variance of $\sqrt{n}\,\hat{\lambda}_{J_0}$, in the limit, can be given by:

$$V(\sqrt{n}\ \hat{\lambda}_{J_0}) = \sigma_y^2\ [\log_{n\rightarrow\infty} \beta'X'\ A_z\ M_x\ A_z\ X\beta / n]^{-1}$$

On the other hand, the estimated variance $\sqrt{n}\ \hat{\lambda}_{J_0}$ from the compound regression model H_J can be written as:

$$\hat{V}(\sqrt{n}\ \hat{\lambda}_{J_0}) = \hat{\sigma}_J^2\ [y'A_z M_x A_z y / n]^{-1},$$

where $\hat{\sigma}_J^2 = (y'\ M_J\ y / n)$, $M_J = I_n - X_J\ (X_J{}'X_J)^{-1}\ X_J'$, and $X_J = [X \mid Z\hat{\gamma}]$. Using partition inversion it is possible to write $y'\ M_J\ y$ as:

$$y'\ M_J\ y = y'\ M_X\ y - \frac{[y'A_Z M_X\ y]^2}{y'A_Z M_X A_Z y}$$

Substituting $y = X\beta + u_0 - e$ and dropping terms of lower order in probability implies that $\text{plim}_{H_0}\ (y'\ M_J\ y / n) = \sigma_y^2$, since $\text{plim}_{H_0}\ (y'A_Z M_X A_Z y / n) = \log_{n\rightarrow\infty} [\beta'X'\ A_z\ M_x\ A_z\ X\beta / n]$, these variances are equivalent and the J-test is a correct test of H_0 even if y is mismeasured.

In conclusion, measurement errors in the regress and does not alter the asymptotic behavior of least-squares-based separate tests if the competing models consist of fixed or non-stochastic regressors. On the other hand, an EVM with mismeasured regressors will always be rejected asymptotically regardless of whether it is “true” or not. These test statistics for H_0 converge to non-central limits when H_0 contains mismeasured regressors, implying a non-conformity between the nominal and true sizes of these tests. Thus, to draw correct inferences in such situations, several modified forms of these tests are proposed in the next section.

4.4. IV Estimation of EVM and Modified Non-Nested Tests

Asymptotic divergence between the nominal and true sizes of conventional non-nested tests of a mismeasured H_0 is attributed to the inconsistent least squares estimation of an EVM. One method of obtaining consistent estimators for mismeasured models is IV estimation. Thus, one logical solution to the asymptotic distortions of these separate tests may simply involve the use of IV estimators for mismeasured models instead of least squares estimators.

In this section, H_0 is initially assumed to be the EVM; IV estimators of H_0 are derived to modify various non-nested tests of H_0. The asymptotic behavior of modified forms of the linearized Cox test, J-, JA-tests, and encompassing test are evaluated in separate subsections.

To derive the IV estimators, $\tilde{\beta}$ and $\tilde{\sigma}_0^2$, consider the existence of an nxk matrix, W, of instrumental variables for the matrix X containing mismeasured regressors. Here, W consists of at least as many columns as in X, i.e., $k \geq k_0$, and correctly measured or non-stochastic regressors in X may also appear in W. In addition, W satisfies the following assumptions.

(i) $\text{plim} \dfrac{W'X}{n} = Q_{WX}$, $Q'_{WX} = Q_{XW}$, a finite, and non-singular matrix.

(ii) $\text{plim} \dfrac{W'W}{n} = Q_{WW}$, a finite, non-singular, and positive definite matrix.

(iii) $\text{plim} \dfrac{W'u}{n} = 0$, a k-dimension null vector.

The first two assumptions (i) and (ii) ensure the existence of IV estimators and the third establishes their consistency.

The generalized IV estimators, $\tilde{\beta}$ and $\tilde{\sigma}_0^2$, may be now defined as:

$$\tilde{\beta} = (\hat{X}'\hat{X})^{-1}\hat{X}'y \tag{4.21}$$

$$\tilde{\sigma}_0^2 = \frac{\tilde{u}_0'\tilde{u}_0'}{n} \tag{4.22}$$

where $\hat{X} = W(W'W)^{-1}W'X$, a matrix of predicted values obtained by regressing each column of X on the entire matrix W, and $\hat{u}_0 = (y - X\tilde{\beta})$, an IV-based residual vector of the model H_0. The consistency of these IV estimators can be readily demonstrated as:

$$\text{plim } \tilde{\beta} = \text{plim } (\hat{X}'\hat{X} / n)^{-1} \text{ plim } [\hat{X}'(X\beta + u_0) / n]$$
$$= \beta + (Q_{xw} Q_{ww}^{-1} Q_{wx})^{-1} Q_{xw} Q_{ww}^{-1} . 0 = \beta$$
$$\text{plim } \tilde{\sigma}_0^2 = \text{plim } [(X\beta + u_0 - X\tilde{\beta})'(X\beta + u_0 - X\tilde{\beta}) / n]$$
$$= \text{plim } (u_0' u_0 / n) = \sigma_0^2$$

In the discussion below, the linearised Cox, J-, JA-, and WET statistics are subjected to modifications based on IV estimators and asymptotic evaluation under the mismeasured H_0. However, these findings will be generalized to other separate tests.

4.4.1. Linearized Cox Test

Replacing the least squares estimators, $\hat{\beta}$ and $\hat{\sigma}_0^2$, with their IV-counterparts, $\tilde{\beta}$ and $\tilde{\sigma}_0^2$ the linearized Cox test in equation (4.7) can be expressed in its modified form as:

$$S_0^m = -(\sqrt{n} / 2)([\tilde{\sigma}_{10}^2 - \hat{\sigma}_1^2] / \hat{\sigma}_1^2) \qquad (4.23)$$

where

$$\tilde{\sigma}_{10}^2 = \tilde{\sigma}_0^2 + (1/n)\, \tilde{\beta}'X'M_z X\tilde{\beta} \qquad (4.24)$$

In the following theorem, however, it is shown that such a simple modification is not sufficient to obtain an asymptotically non-zero mean for S_0^m under the mismeasured H_0.

THEOREM 4.4

Under the mismeasured H_0 the modified statistic does not have a zero mean asymptotically.

PROOF

From equation (4.11):

$$\text{plim}_{H_0} = \sigma_0^2 + \beta'(Q_{xx} - Q_{xz} Q_{zz}^{-1} Q_{zx})\beta - 2\beta'\beta \neq 0$$

$$\text{plim}_{H_0}\ \tilde{\sigma}_{10}^2 = \sigma_0^2 + \beta' (Q_{xx} - Q_{xz} Q_{zz}^{-1} Q_{zx})\ \beta \neq 0$$

Thus,

$$\text{plim}_{H_0}\ S_0^m \frac{-\sqrt{n}\beta' Q_{V_0} \beta}{\sigma_0^2 + \beta'(Q_{xx} - Q_{xz} Q_{zz}^{-1} Q_{zx})\beta - 2\beta' Q_{V_0} \beta} \neq 0$$

Another problem with the above modification may be its non-conformity with the first principle of Cox in which $\tilde{\sigma}_{10}^2$ is to be obtained by evaluating unknown parameters in $\text{plim}_{H_0}\ \hat{\sigma}_1^2$ at their consistent estimators. However, the following theorem indicates that this second modification to the Cox test will not exist.

THEOREM 4.5

The linearized Cox test for mismeasured H_0 using the IV estimators of H_0 will not be identified.

PROOF

In the proof to Theorem 4.4, it was shown that

$$\text{plim}_{H_0}\ \hat{\sigma}_1^2 = +\ \beta' (Q_{xx} - Q_{xz} Q_{zz}^{-1} Q_{zx})\ \beta - 2\beta'\ Q_{V_0}\ \beta \neq 0$$

Here β and σ_0^2 may well be replaced by their consistent IV estimators, but is, in general, not identified.

Following theorems 4.4 and 4.5, limitation of the modified linearized Cox test for testing a mismeasured H_0 can be analogously extended to the variance encompassing test of Mizon (1984) and G-test of Godfrey (1983) which will also involve $\hat{\sigma}_1^2$ and $\hat{\sigma}_{10}^2$. On the other hand, IV estimators are used in the following to modify the J- and JA-tests and it is shown that such modifications will lead to asymptotically valid and correct tests for a mismeasured H_0.

4.4.2. J- and JA-Tests

Substituting the IV estimators of H_0 for its least squares estimators in equations (4.8) and (4.9) may lead to the following IV-based expressions for the J- and JA-test statistics respectively:

$$\sqrt{n}\,\tilde{\lambda}_{J_0} = \sqrt{n}\frac{\hat{y}'_1 M_x y}{\hat{y}'_1 M_x M'_x \hat{y}_1} \tag{4.25}$$

$$\sqrt{n} = \frac{\hat{y}'_{10} M_x y \sqrt{n}\tilde{\lambda}_{J_0}}{\tilde{y}'_{10} M_x M'_x \tilde{y}_{10}} \tag{4.26}$$

where $M_x = [I_n - X(\hat{X}'\hat{X})^{-1}\hat{X}']$ is an IV projection matrix for H_0; $\hat{y}_1 = A_z y$ and $\tilde{y}_{10} = A_z \tilde{y}_0$ are least squares, vectors of predicted values obtained from regressions of y and $\tilde{y}_0$ on Z respectively. Here $\hat{y}_0$ represents IV-based predicted values from H_0. From here on, these tests will be termed the IV-based J- and JA-tests and their asymptotic behavior is studied in the following theorem.

THEOREM 4.6

The IV-based statistics $\sqrt{n}\tilde{\lambda}_{J_0}$ and $V\sqrt{n}\tilde{\lambda}_{J_0}$ are asymptotically normal with zero mean under the mismeasured H_0 and are asymptotically unbounded under H_1.

PROOF

Replacing y with $X\beta + u_0$ in (4.25) and retaining terms of the highest order in probability yields:

$$\sqrt{n}\tilde{\lambda}_{J_0} = \frac{\beta' X' A_z \tilde{M}_x u_0}{\beta' X' A_z \tilde{M}_x \tilde{M}'_x A_z X\beta}$$

From the normality assumption of u_0, i.e., $u_0 \sim N(0, \sigma_0^2 I_n)$, and Lindberg-Feller's central limit theorem, it follows that $\sqrt{n}\tilde{\lambda}_{J_0}$ is asymptotically normal variate with zero mean and variance defined as follows:

$$V\sqrt{n}\tilde{\lambda}_{J_0} = n\,\sigma_0^2\,\frac{\beta' X' A_Z \tilde{M}_x \tilde{M}'_x A_Z X\beta}{(\beta' X' A_Z \tilde{M}_x \tilde{M}'_x A_Z X\beta)^2} \tag{4.27}$$

Thus, a standard normal test for H_0 may be given by:

$$\tilde{J}_0 = \frac{\sqrt{n}\; y'_1 M_X y}{\tilde{\sigma}_0 (n\; y'_1 M_X M'_x y_1)^{1/2}} \tag{4.28}$$

Similarly, JA_0 can be obtained by replacing $\hat{y}_1$ with $\hat{y}_{10}$ in (4.27) and the above calculations may be modified to show that JA_0 is asymptotically standard normal under H_0

Next, the asymptotic behavior of these tests under H_1 is characterized by:

$$\text{plim}_{H_1}(\sqrt{n}\,\tilde{\lambda}_{J_0}) = \sqrt{n}\,\frac{\text{plim}_{H_1}(y'_1 M_x y / n)}{\text{plim}_{H_1}(y'_1 M_x M'_x y_1 / n)}$$

$$= \sqrt{n}\,\frac{\text{plim}(\gamma' Z' M_X Z\gamma / n)}{\text{plim}(\gamma' Z' M_x M'_x Z\gamma / n)} \tag{4.29}$$

the latter term being O_p (1/2). Similarly, $\text{plim}_{H_1}(\sqrt{n}\,\hat{\lambda}_{JA_0}) = O_p(1/2)$. Thus, both of these statistics are asymptotically unbounded under H_1.

Although these IV-based tests are asymptotically well behaved, their interpretation and computation as the t-ratios from augmented regression models is not possible. Therefore, using the formal approach of artificial regression models, a modified J-test for mismeasured H_0 may be now obtained as a t-ratio for the IV estimators of λ_{J_0} in the following artificial regression model:

$$H_{J_0}: y = X\beta (1 - \lambda_{J_0}) + \hat{y}_1 \lambda_{J_0} + u_{J_0},$$

where $\hat{y}_1 = Z\hat{\gamma}$. Assuming = [W | $\hat{y}_1$] as an IV matrix for = [X | $\hat{y}_1$], the IV estimator of λ_{J_0} may be written as:

$$\sqrt{n}\,\tilde{\lambda}_{J_0^m} = \tag{4.30}$$

$$\sqrt{n}\ \frac{y'_1 (I_n - X[\hat{X}'\hat{X} + e'_w y_1 (y'_1 M_w y_1)^{-1} y'_1 e_w]^{-1} [\hat{X}' + e'_w y_1 (y_1 M_w y_1)^{-1} y'_1 M_w) y}{y'_1 (I_n - X[\hat{X}'\hat{X} + e'_w y_1 (y'_1 M_w y_1)^{-1} y'_1 e_w]^{-1} X') y_1}$$

where $\hat{X} = A_w X$
$e_w = M_w X$
$A_w = W(W'W)^{-1} W'$
$M_w = I_n - A_w.$

The estimated variance of $\sqrt{n}\,\tilde{\lambda}_{J0}^{m}$, denoted as n V $\sqrt{n}$ ($\tilde{\lambda}_{J0}^{m}$), may be obtained as the product of n $\tilde{\sigma}^2_{J_0^m}$ and inverse of the denominator term in equation (4.30).

Here, $\tilde{\sigma}^2_{J_0^m} = (\tilde{u}'_{J_0} \tilde{u}_{J_0} / n)$ where $\tilde{u}_{J_0}$ is the IV residual vector from artificial regression model H_{J_0}. Thus, the modified J-test when H_0 is mismeasured may be defined as the t-ratio of $\tilde{\lambda}_{J0}^{m}$ from IV estimation of H_{J_0}, i.e.,

$$\tilde{J}_0^m = (\sqrt{n}\,\tilde{\lambda}_{J0}^{\;m} / [\mathrm{n\,V}(\tilde{\lambda}_{J0}^{\;m})]^{1/2}). \tag{4.31}$$

The modified JA-test is similarly obtained from an IV estimation of the following artificial regression model:

$$\mathrm{H}_{\mathrm{JA}_0}: \mathrm{y} = \mathrm{X}\beta\,(1 - \lambda_{JA_0}) + \hat{\mathrm{y}}_{10}\,\lambda + u_{JA_0},$$

where $\tilde{\mathrm{y}}_{10} = \mathrm{Z(Z'Z)}^{-1}Z'X\,\beta$ is a vector of least squares predicted values when IV predicted values of H_0 are regressed on Z matrix. The modified JA-test for mismeasured H_0, denoted as $\tilde{\mathrm{JA}}_0^{\mathrm{m}}$, can be directly obtained from (4.30) by replacing $\hat{\mathrm{y}}_1$ with $\tilde{\mathrm{y}}_{10}$.

At this point, some clarification about the notation used in the remainder of this book may be useful. The IV-based J- and JA-tests, denoted earlier as J_0 and JA_0, are developed by simple replacement of least squares estimators with IV estimators (and may lack practical appeal due to computational difficulties, since they cannot be readily obtained from an artificially embedded regression model). On the other hand, the tests obtained as t-ratios from the IV estimation of $\mathrm{H}_{\mathrm{J}_0}$ and $\mathrm{H}_{\mathrm{JA}_0}$ above, termed the modified IV tests and denoted $\tilde{J}_0^m$ and $\tilde{\mathrm{JA}}_0^{\mathrm{m}}$, providing relative ease of computation, will be the focus of the remaining analysis.

THEOREM 4.7

Under a mismeasured H_0, the modified J- and JA-tests, $\tilde{J}_0^m$ and $\tilde{\mathrm{JA}}_0^{\mathrm{m}}$, are asymptotically normal with zero mean, have the correct size, and are asymptotically equivalent. Moreover, these are consistent statistics.

PROOF

Substituting $\mathrm{X}\beta + \mathrm{u}_0$ for y in the numerator and denominator of equation (4.30) and retaining terms of the highest order in probability yields the following expressions respectively:

$$NUM_{H_0}(\sqrt{n}\,\tilde{\lambda}_0^m) =$$

$$\sqrt{n}\beta' X' A_Z \left(I_n - X\left(\hat{X}'\hat{X} + \frac{e'_W A_Z X\beta\beta' X' A_Z e_w}{\beta'\mathrm{X'A_Z M_W A_Z}X\beta}\right)^{-1}\left(\hat{\mathrm{X}}' + \frac{e'_W A_Z X\beta\beta' X' A_Z M_w}{\beta'\mathrm{X'A_Z M_W A_Z}X\beta}\right)\right)\mathrm{u}_0 \tag{4.32}$$

$$DEN_{H_0}(\sqrt{n}\,\tilde{\lambda}_{J0}^{m}) =$$

$$\beta' X' A_Z \left(I_n - X\left(\hat{X} + \frac{e'_W A_Z X\beta\beta' X' A_Z e_w}{\beta' X' A_Z M_W A_Z X\beta}\right)^{-1}(X)\right) A_Z X\beta. \tag{4.33}$$

Invoking normality of u_0, i.e., $u_0 \sim N(0, \sigma_0^2 I_n)$ and applying Lindberg-Feller's central limit theorem, it can be established that $NUM_{H_0}(\sqrt{n}\,\tilde{\lambda}_{J0}^{m})$ is asymptotically normal variate with zero mean and variance given by (after some manipulations from (4.32):

$$V\left[NUM_{H_0}(\sqrt{n}\,\tilde{\lambda}_{J0}^{m})\right] = \tag{4.34}$$

$$n\sigma_0^2 \operatorname{plim} \beta' X' A_z \left(I_n - X\left(\hat{X}'\hat{X}' + \frac{e_w' A_z X\beta\beta' X' A_z e_w}{\beta' X' A_z M_w A_z X\beta}\right)^{-1} X'\right) A_z X\beta$$

$$= n\,\sigma_0^2 \operatorname{plim} DEN_{H_0}(\sqrt{n}\,\tilde{\lambda}_{J0}^{m})$$

Thus, the statistic $\sqrt{n}\,\tilde{\lambda}_{J0}^{m} \sim N[0, n\,V(\tilde{\lambda}_{J0}^{m})]$where $n\,V(\tilde{\lambda}_{J0}^{m})$ can be easily obtained from (4.33) and (4.34) as:

$$n\,V(\tilde{\lambda}_{J0}^{m}) = \frac{V(NUM_{H_0}(\sqrt{n}\tilde{\lambda}_{Jo}^{m}))}{(\operatorname{plim} DEM_{H_0}(\sqrt{n}\tilde{\lambda}_{Jo}^{m}))^2}$$

$$= \sigma_0^2 (\operatorname{plim}(\sqrt{n}\,\tilde{\lambda}_{J0}^{m}) / n)^{-1} \tag{4.35}$$

The equivalence between the true asymptotic size and nominal size of $\sqrt{n}\,\tilde{\lambda}_{J0}^{m}$ follows from

$$\operatorname{plim}_{H_0}[n\,V(\tilde{\lambda}_{J0}^{m})] = \operatorname{plim}_{H_0} \tilde{\sigma}_{J_0}^{2m} (\operatorname{plim} DEN_{H_0}(\sqrt{n}\,\tilde{\lambda}_{J0}^{m}) / n)^{-1}$$

$$= n\,V(\tilde{\lambda}_{J0}^{m}) \tag{4.36}$$

since $\text{plim}_{H_0} \tilde{\sigma}_{J_0}^{2m} = \text{plim}_{H_0} (u'_0 u_0 / n) \sigma_0^2$. Here $\tilde{J}_0^m$ is an asymptotically standard normal statistic.

Similarly, following the above steps, asymptotic normality and correct size of the modified JA-test can be established; these proofs are not repeated here. Moreover, the mean and variance of $\sqrt{n}\tilde{\lambda}_{JA_0}^m$ collapse to the same limits as that of $\sqrt{n}\tilde{\lambda}_{J_0}^m$. Thus, both statistics are asymptotically equivalent under a mismeasured H_0.

Next, to prove consistency of the modified J-test, consider the limiting behavior of under H_1 as follows:

$$\text{plim}_{H_1} \text{NUM}(\sqrt{n}\tilde{\lambda}_{J_0}^m) =$$

$$\sqrt{n}\text{plim}\frac{1}{n}\gamma' Z'\left(I_n - X\left(\hat{X}'\hat{X}' + \frac{e_w' Z\gamma\gamma' Z' e_w}{\gamma' Z' M_w Z\gamma} \right)^{-1} \left(\hat{X}' + \frac{e_w' Z\gamma\gamma' Z' M_w}{\gamma' Z' M_w Z\gamma} \right) \right) Z\gamma, \tag{4.37}$$

$$\text{plim}_{H_1} \text{DEN}(\sqrt{n}\tilde{\lambda}_{J_0}^m) =$$

$$\text{plim}(1/n\,\gamma' Z'\left(X\left(\hat{X}'\hat{X}' + \frac{e_w' Z\gamma\gamma' Z' e_w}{\gamma' Z' M_w Z\gamma} \right)^{-1} X' \right) Z\gamma \tag{4.38}$$

implying

$$= \frac{\text{plim}_{H_1} \text{NUM}(\sqrt{n}\tilde{\lambda}_{J0}^m)}{\text{plim}_{H_1} \text{DEN}(\sqrt{n}\tilde{\lambda}_{J0}^m)} = \sqrt{n}. \tag{4.39}$$

Thus, noting that $\text{plim}_{H_1} n \, V(\tilde{\lambda}_{J0}^m) > 0$. $\tilde{J}_0^m$ is an unbounded statistic under H_1 and will reject a false H_0 with unit probability. Similarly, $\text{plim}_{H_1} \sqrt{n}\tilde{\lambda}_{JA_0}^m = \sqrt{n}$. Hence, both $\tilde{J}_0^m$ and $\tilde{\lambda}_{JA_0}^m$ are consistent statistics.

The computation of modified J- and JA-tests of H_1 (where H_1 is correctly measured) can be based on the least squares estimators of λ from the following artificial regression models respectively:

$H_{J_1}: y = Z\gamma\,(1-\lambda_{J_1}) + y_0\,\lambda_{J_1} + u_{J_1}$

$H_{JA_1}: y = Z\gamma\,(1-\lambda_{JA_1}) + y_{01}\,\lambda_{JA_1} + u_{JA_1}$

Here $y_0 = X\tilde{\beta}$

$\tilde{y}_{01} = X(\hat{X}'\hat{X})^{-1}\hat{X}'\hat{y}_1$

$\hat{y}_1 = Z(Z'Z)^{-1}Z'y$

In this case,

$$\sqrt{n}\,\hat{\lambda}_{J_1} = \sqrt{n}\,[\tilde{y}'_0 M_z y / \tilde{y}'_0 M_z \tilde{y}_0] \qquad (4.40)$$

$$\sqrt{n}\,\hat{\lambda}_{JA_1} = \sqrt{n}\,[\tilde{y}'_{01} M_z y / \tilde{y}_{01} M_z \tilde{y}_{01}] \qquad (4.41)$$

These statistics are also asymptotically normal with zero mean and have the correct size asymptotically when H_1 is the true model. Moreover, they will be consistent tests of H_1. A modified test based on encompassing principle, which may also be interpreted as the F- or R-test based on IV estimation of a comprehensive model, is now discussed.

4.4.3. Encompassing Test

A modified form of the WET statistic, $\tilde{\eta}^m_{w_0}$ when H_0 is mismeasured against a correctly measured H_1, involving the slope parameter vector $\hat{\gamma}_1$ (the sub-vector $\hat{\gamma}$ of associated with submatrix, $\overline{Z}$, of distinct regressors in H_1), can be defined as:

$$\tilde{\eta}^m_{w_0} = \frac{n(\hat{\gamma}_1 - \hat{\gamma}_{10})'(\hat{\gamma}_1 - \hat{\gamma}_{10})}{\tilde{V}(\hat{\gamma}_1 - \hat{\gamma}_{10})} \qquad (4.42)$$

where $\hat{\gamma}_1 = (\overline{Z}'\overline{Z})^{-1}\overline{Z}'y$

$$\tilde{\gamma}_{10} = (\text{plim}_{H_0}\ \hat{\gamma}_1)|_{\beta = \tilde{\beta}} = (\overline{Z}'\ \overline{Z})^{-1}\ \overline{Z}'X\ \tilde{\beta}$$

$$V(\hat{\gamma}_1 - \tilde{\gamma}_{10}) = n\ \tilde{\sigma}_0^2\ (\overline{Z}'\ \overline{Z})^{-1}\ \overline{Z}'M_x\ M'_x\ (\overline{Z}'\ \overline{Z})^{-1} \quad (4.43)$$

Here, $M_x = [I_n - X(\hat{X}'\ \hat{X})^{-1}\ \hat{X}']$, $\tilde{\sigma}_0^2 = (u'_c\ u_c / n)$ and u_c represents the IV residual vector from the compound model: $y = X\beta + \overline{Z}\ \gamma_1 + u_c$. An equivalent expression for $\hat{\eta}_{w_0}^m$ may be given by:

$$\hat{\eta}_{w_0}^m = \tilde{\sigma}_0^{-2}\ u'_0\ \overline{Z}\ (\overline{Z}'M_x\ M'_x\ \overline{Z})^{-1}\ \overline{Z}'\ u_0 \quad (4.44)$$

The following theorem will now describe the asymptotic behavior of this test, $\tilde{\eta}_{w_0}^m$.

THEOREM 4.8

When H_0 contains mismeasured repressors, the modified WET statistic, $\hat{\eta}_{w_0}^m$, is an asymptotically central $X_{k_1}^2$ - r is a consistent test.

PROOF

Consider $\sqrt{n}(\hat{\gamma}_1 - \hat{\gamma}_{10}) = \sqrt{n}\ (\overline{Z}'\ \overline{Z})^{-1}\ \overline{Z}'\ M_x\ y$. Substituting $X\beta + u_0$ for y and retaining terms of the highest order in probability yields the following:

$$\sqrt{n}(\hat{\gamma}_1 - \hat{\gamma}_{10}) = \sqrt{n}\ (\overline{Z}'\ \overline{Z})^{-1}\ \overline{Z}'\ M_x\ u_0$$

Invoking the normality assumption of u_0, $\sqrt{n}(\hat{\gamma}_1 - \tilde{\gamma}_{10})$ represents a vector of zero mean normal variates with asymptotic variance-covariance matrix given by:

$$n\ V(\hat{\gamma}_1 - \tilde{\gamma}_{10}) = n\ \sigma_0^2\ \text{plim}\ (\overline{Z}'\ \overline{Z})^{-1}\ \overline{Z}'\ M_x\ M'_x (\overline{Z}'\ \overline{Z})^{-1}$$

Noting that, $\text{plim}\,\tilde{\sigma}_0^2 = \sigma_0^2$ establishes that $\text{plim}_{H_0}\,\tilde{V}(\hat{\gamma}_1 - \tilde{\gamma}_{10}) = n\,V(\hat{\gamma}_1 - \tilde{\gamma}_{10})$. Thus, $\tilde{\eta}_{w_0}^m$ is an asymptotically central $\chi_{K_1}^2$ - r.

Consistency of $\tilde{\eta}_{w_0}^m$ can he proved by noting that

$$\text{plim}_{H_1}\,\sqrt{n}\,(\hat{\gamma}_1 - \tilde{\gamma}_{10}) = \sqrt{n}\;\text{plim}_{H_1}\,(\overline{Z}'\,\overline{Z})^{-1}\overline{Z}'\,M_x\,(Z\gamma + u_1)$$
$$= \sqrt{n}\,(\gamma_1 - \overline{Q}_{zz}^{\,-1}\,\overline{Q}_{zx}\,\hat{Q}_{xx}^{\,-1}\,\hat{Q}_{xz}\,\gamma_1) \neq 0,$$

where $\overline{Q}_{zz} = \lim_{n\to\infty}(\overline{Z}'\,\overline{Z}/n)$, $\overline{Q}_{zx} = \text{plim}\,(\overline{Z}'X/n)$, $\hat{Q}_{xx} = \text{plim}\,(\hat{X}'\hat{X}/n)$, and $\hat{Q}_{xz} = \text{plim}\,(\hat{X}'Z/n)$ are finite and non-zero. Thus, $\hat{\eta}_{w_0}^m$ will be an asymptotically unbounded when H_1 is the true model.

The slope parametric encompassing test and Dastoor's (1983) R-Lest are numerically equivalent, therefore, the above results of Theorem 4.7 will also be valid for Dastoor's R-test. However, the variance encompassing statistic, $\sqrt{n}(\hat{\sigma}_1^2 - \tilde{\sigma}_{10}^2)$, cannot be obtained similarly since $\tilde{\sigma}_{10}^2$ is shown to be unidentified in Theorem 4.5. The conventional F-test for $\gamma_1 = 0$, derived from IV estimation of the following compound model: $y = X\beta + \overline{Z}\gamma_1 + u_c$, will also be consistent and asymptotically equivalent to $\hat{\eta}_{w_0}^m$ under a mismeasured H_0.

The next section extends the analysis to allow for the possibility of both H_0 and H_1 containing mismeasured regressors. Before concluding the current section, a brief discussion of the appropriate specification of the instrument set, W, is provided. So far the existence of W has been presumed subject only to a set of mathematical and statistical properties. Recalling that H_0 is an EVM while H_1 is a correctly measured regression model satisfying the full ideal conditions, the IV conditions involve the existence and non-singularity Q_{xz} and Q_{zz}, as well as asymptotic non-correlation of Z and u_o. The following theorem illustrates that careful consideration must be given to the selection of W. In particular, although Z, the correctly measured regressor set in H_1 satisfies the IV conditions, and even though it may contain completely distinct columns from X, it may not be used as the instrument set.

THEOREM 4.9

The modified J-statistic $\sqrt{n}\ \tilde{\lambda}_{J0}^{m}$ defined in (4.30) collapses to a constant when Z is used as an IV matrix for the mismeasured matrix X.

PROOF

Replacing A_w and M_w with A_z and M_z respectively in (4.30) and noting that $Z'A_z = Z$ and $Z'M_z = 0$ yields the following:

$$\tilde{\lambda}_{J_0} = \frac{y' A_Z \tilde{M}_X A_Z y}{y' A_Z \tilde{M}_X A_Z y} = 1$$

Thus, irrespective of H_0 or H_1 being a true model, $\tilde{\lambda}_{J_0}$ equals 1.

It can be similarly shown that above result will apply when, for instance, each column in W is a linear combination of some or all of the columns in Z. Analogously, these results can be extended to other modified tests such as the JA-test and encompassing test. Thus, testing mismeasured models within a non-nested framework involves additional restrictions regarding linear dependencies between the IV matrix for one model and regressors of the rival model.

The main purpose of this section is to analyze certain properties of various non-nested tests when H_0 is mismeasured. The major results can be summarized as follows: the presence of mismeasured regressors in H_0 does not affect the consistency property of the conventional non-nested tests while it renders their asymptotic sizes to be invalid under H_0; the presence of mismeasured regressands only in non-nested models, however, does not affect the asymptotic properties of the conventional separate tests; the linearized Cox test cannot be modified by replacing least squares with IV estimators, but modified forms for the J-, JA-, and encompassing tests are consistent and have correct asymptotic size under the mismeasured H_0; in applying the modified J-, JA-, or encompassing test, the choice of an IV matrix, W, for mismeasured models requires no linear dependence between the columns of Z and W.

The next section will extend the role of the modified tests when both H_0 and H_1 are mismeasured and will evaluate the asymptotic power of these tests under the fixed and Pesaran (1982) type local alternative models.

4.5. New Tests when Both Non-Nested Models are Mismeasured

The earlier sections restricted the specification of H_0 to be an EVM whereas the (non-nested) alternative H_1 was a completely observed and correctly measured model, a situation that may be implausible in practice. Alternatively, an incorrectly specified H_1 (due to the presence of mismeasured regressors) may undermine its role as a serious contender to explain real world economic phenomena. The following discussion extends the earlier analysis of the modified tests for a mismeasured H_0 accompanied with an EVM H_1. Formally, the two non-nested models will follow the EVM specification used in the previous section:

$$H_0: y = X\beta + u_0$$
$$H_1: y = Z\gamma + u_1$$

Earlier assumptions in section 4.2 are maintained except that the Z matrix now consists of some or all mismeasured regressors so that $u_1 = \varepsilon - V_1 \gamma$. Here, ε is an nxl vector of structural, stochastic disturbance terms in H_1 and V_1 is an nxk_1 matrix of measurement errors in Z whereas the rank of V_1 will depend on the number of mismeasured regressors in Z. It is assumed that $\varepsilon \sim N(0, \sigma_{\varepsilon}^2 I_n)$, $V_{i_1} \sim N(0, \sigma_{V_1}^2 I_n)$, and $E(\varepsilon' V_{i_1}) = 0$ for each column vector in V_1. Thus, plim $(u_1'u_1 / n) = \sigma_1^2$ and plim $(Z'u_1 / n) = -Q_{V_1}\gamma$ where $\sigma_1^2 = \sigma_{\varepsilon}^2 + \gamma' Q_{V_1} \gamma$ and Q_{V_1} = plim $(V_1'V_1 / n)$. If X and Z consist of distinct mismeasured regressors and assuming that the measurement errors in both matrices are asymptotically uncorrelated, i.e., plim $(V'_0 V_1 / n) = 0$, then it may be fair to assume that plim $(Z' u_0 / n)$ = plim $(X' u_1 / n) = 0$; otherwise, plim $(Z'u_0 / n)$ and plim $(X'u_1 / n)$ may be respectively represented by non-zero matrices denoted as $Q_{V_{10}}$ and $Q_{V_{01}}$. Moreover, assuming plim $(W'Z / n) = Q_{wz}$ to be a finite, non-singular matrix and plim $(W'u_1 / n)$ a null vector, the IV estimators for γ and σ^2_1 can be defined as follows:

$$\tilde{\gamma} = (\hat{Z}'\hat{Z})^{-1} \hat{Z} y$$
$$\tilde{\sigma}_1^2 = \tilde{u}'_1 \tilde{u}_1 / n$$

where $\hat{Z} = W(W'W)^{-1} W'Z$ and $u_1 = y - Z\tilde{\gamma}$. When H_1 is now mismeasured, most of the earlier results regarding the asymptotic behavior of non-nested tests hold true. For brevity, then, this section will review these properties when H_1 is mismeasured while providing necessary alterations in the equations to accommodate mismeasurement in H_1.

The choice of a null or alternative model in non-nested testing is arbitrary and both models assume these roles alternately. Therefore, following Theorem 4.1 the conventional non-nested tests for mismeasured H_1 will cause its rejection whether H_1 is "true" or not. The results in theorems 4.4 and 4.5, non-zero mean and non-existence of the modified Cox test of a mismeasured model, immediately follow when $\text{plim}_{H_0} \hat{\sigma}_1^2$ is replaced with the following expression:

$$\text{plim}_{H_0} \tilde{\sigma}_1^2 = \sigma_0^2 - 2\,\beta' Q_{V_0}\,\beta + \beta' Q_{xx}\,\beta - 2\,\text{plim}\,(u'_0 Z(\hat{Z}'\hat{Z})^{-1}\hat{Z}'X\beta/n) - 2\,\text{plim}\,(\beta'X'Z(\hat{Z}'\hat{Z})^{-1}\hat{Z}'X\beta/n) + \text{plim}\,(\beta'X'(\hat{Z}'\hat{Z})^{-1}Z'Z(\hat{Z}'\hat{Z})^{-1}\hat{Z}\,X\beta/n). \tag{4.45}$$

The IV-based J- and JA-tests, J_0 and JA_0, can similarly be extended when H_1 is also mismeasured. Replacing least squares predicted values of H_1, $\hat{y}_1$, with IV predicted values of H_1, y_1, in equation (4.28) yields the following standard normal statistic:

$$\tilde{J}_0 = \frac{\sqrt{n}\,\tilde{y}_1'\tilde{M}_X y}{\tilde{\sigma}_0 (n\tilde{y}_1'\tilde{M}_X\tilde{M}'_X\tilde{y}_1)^{1/2}}. \tag{4.46}$$

Similarly, $J\tilde{A}_0$ can be defined in the following by replacing $\tilde{y}_1$ in (4.46) with $\tilde{\tilde{y}}_{10}$. Here, $\tilde{\tilde{y}}_{10}$ is an IV predicted vector which is obtained by regressing on the matrix Z.

$$J\tilde{A}_0 = \frac{\sqrt{n}\,\tilde{\tilde{y}}'_{10}\tilde{M}_X y}{\tilde{\sigma}_0 (n\tilde{\tilde{y}}'_{10}\tilde{M}_X\tilde{M}'_X\tilde{\tilde{y}}_{10})^{1/2}}. \tag{4.47}$$

The modified J- and JA-tests can be respectively obtained as IV estimated t-ratios from the following artificial regression models:

$$H_{J_0}: y = X\beta\,(1 - \lambda_{J_0}) + \tilde{y}_1\,\lambda_{J_0} + u_{J_0}$$

$$H_{JA_0}: y = X\beta\,(1 - \lambda_{JA_0}) + \tilde{\tilde{y}}_{10}\,\lambda_{JA_0} + u_{JA_0}$$

The J-test of a mismeasured H_0 from IV estimation of H_{J_0} may be given by:

$$\tilde{J}_0^{m} = \frac{\sqrt{n}\tilde{y}'_1\left(I_n - X[X'A_{W_{J_0}}X]^{-1}X'A_{W_{J_0}}\right)y}{\tilde{\sigma}_{J_0}[n\tilde{y}'_1\left(I_n - X[X'A_{W_{J_0}}X]^{-1}X'\right)\tilde{y}_1]^{1/2}}. \qquad (4.48)$$

$$J\tilde{A}_0^{m} = \frac{\sqrt{n}\tilde{\tilde{y}}'_{10}\left(I_n - X[X'A_{W_{JA_0}}X]^{-1}X'A_{W_{JA_0}}\right)y}{\tilde{\sigma}_{JA_0}[n\tilde{\tilde{y}}'_{10}\left(I_n - X[X'A_{W_{JA_0}}X]^{-1}X'\right)\tilde{\tilde{y}}_{10}]^{1/2}}. \qquad (4.49)$$

Here $Aw_{J_0} = W_{J_0}\,(W'_{J_0}\,W_{J_0})^{-1}W'_{J_0}$ and $Aw_{JA_0} = W_{JA_0}\,(W'_{JA_0}\,W_{JA_0})^{-1}W'_{JA_0}$ whereas $W_{J_0} = [W \mid \tilde{y}_1]$ and $W_{JA_0} = [W \mid \tilde{\tilde{y}}_{10}]$ respectively denote IV matrices for H_{J_0} and H_{JA_0}.

MacKinnon et al (1983) and Godfrey (1983), respectively, derive IV based J- and JA-tests which will be numerically equivalent to $\tilde{J}_0^{m}$ and if the same IV matrix, W, for H_0 and H_1 also serves as the IV for H_{J_0} and H_{JA_0}, i.e., $W = W_{J_0}$ and $W = W_{JA_0}$. The formal proofs of asymptotic normality with zero mean under H_0 and consistency of these four tests [(4.46) through (4.49)] will be a direct replication of earlier proofs to Theorems 4.6 and 4.7.

The encompassing test in equation (4.42) can also be rewritten by replacing $\hat{\gamma}_1$ and $\tilde{\gamma}_{10}$ with and respectively:

$$\tilde{\eta}_{w_0}^{m} = \frac{n\,(\tilde{\gamma}_1 - \tilde{\tilde{\gamma}}_{10})'\,(\tilde{\gamma}_1 - \tilde{\tilde{\gamma}}_{10})}{\tilde{V}\,(\tilde{\gamma}_1 - \tilde{\tilde{\gamma}}_{10})} \qquad (4.50)$$

where

$$\tilde{\tilde{\gamma}}_{10} = [\text{plim}_{H_0}\ \tilde{\gamma}_1]|_{\beta=\tilde{\beta}} = (\bar{Z}'A_w\bar{Z})^{-1}\bar{Z}'A_w X\tilde{\beta},$$

$$\tilde{V}(\tilde{\gamma}_1 - \tilde{\tilde{\gamma}}_{10}) = n\,\tilde{\sigma}_0^2(\bar{Z}'A_w\bar{Z})^{-1}A_w M_x M'_x A_w(\bar{Z}'A_w\bar{Z})^{-1}$$

Applying the steps in Theorem 4.8, the statistic $\tilde{\eta}^m_{w_0}$ can be proved an asymptotic central $\chi^2_{K_1} -_r$ under H_0 and also a consistent test.

Although the Cox test for a mismeasured model - involving a comparison between and $\tilde{\sigma}^2_{10}$ - does not exist (Theorem 4.5), Ericsson (1983) and Godfrey (1983) follow Gal-lent and Jorgenson's (1979) IV formulation to develop the following, IV-based Cox test of a mismeasured H_0 :

$$G_0 = \sqrt{n}\,(D_0/n) - d_0)/(\tilde{V}_{G_0})^{1/2} \tag{4.51}$$

where

$$(D_0/n) - d_0 = (\tilde{u}'_1 A_w \tilde{u}_1/n) - (\tilde{u}'_1 A_w \tilde{u}_1/n)\,|_{[\beta=\tilde{\beta},\ \sigma_0^2=\tilde{\sigma}_0^2}$$

$$= y'(\hat{A}_x\hat{A}_z\hat{A}_x - \hat{A}_z)y/n \tag{4.52}$$

$$\tilde{V}_{G_0} = 4\,\tilde{\sigma}_0^2\,[y'\hat{A}_x\hat{M}_z\hat{M}_x\hat{M}_z\hat{A}_x y/n]$$

$$\hat{A}_x = (\hat{X}'\hat{X})^{-1}\hat{X}'$$

$$\hat{A}_z = (\hat{Z}'\hat{Z})^{-1}\hat{Z}'$$

$$\hat{M}_x = I_n - \hat{A}_X$$

$$\hat{M}_z = I_n - \hat{A}_z \tag{4.53}$$

In Ericsson and Godfrey's papers, W serves as the IV matrix for both non-nested models H_0 and H_1. Godfrey (1983, pg. 358-361) shows that G_0 is a standard normal test for H_0 and will be asymptotically equivalent to an IV based JA-type test statistic.

The discussion so far involves the development of various asymptotically valid non-nested tests for one or both mismeasured models. Following the discussion in chapter three, the fixed and local alternative approaches are considered in the following to evaluate asymptotic powers of these IV based tests.

The Fixed Alternative Approach

The fixed alternative approach of Davidson and MacKinnon (1982) is recommended when the null and alternative models are both false; the situation in this study in which both of the models are mismeasured seems to be the best scenario for such an application.

Following Davidson and MacKinnon, the true data generating process, H_T, is specified as:

$$H_T: y = X\beta_0 + ø + u,$$

where β_0 is a fixed parameter, ø is a non-stochastic vector of order $n^{-1/2}$, and $u \sim N(0, \sigma^2 I_n)$. Computation of "fixed alternative'' power requires asymptotic evaluation of a test statistic under H_T. The following theorem now provides an asymptotic comparison of several IV-based non-nested tests under such a fixed alternative.

THEOREM 4.10

The statistics G_0, $\widetilde{J}_0^m$, and $\hat{\eta}_{w_0}^m$ cannot, in general, be ranked in terms of their fixed powers.

PROOF

Returning to the G-test in equations (4.51) through (4.53), substituting y from H_T, and retaining only terms of higher order in probability implies that:

$$G_0 = \frac{\beta'_0 \hat{X}' (\hat{A}_z \hat{A}_x - \hat{A}_z)(ø + u) / \sqrt{n}}{\sigma (\beta'_0 \hat{X}' \hat{M}_z \hat{M}_x \hat{M}_z \hat{X} \beta_0 / n)^{1/2}}$$

As usual, $G_0 \sim N(\mu_{G_0}, 1)$ asymptotically where μ_{G_0} is defined as:

$$\mu_{G_0} = \text{plim} \frac{\beta'_0 \hat{X}' \hat{A}_Z \hat{M}_X ø / \sqrt{n}}{\sigma (\beta'_0 \hat{X}' \hat{M}_z \hat{M}_x \hat{M}_z \hat{X} \beta_0 / n)^{1/2}}$$

Thus, G^2_0 is asymptotically distributed as a non-central X^2_1 with non-centrality parameter given by:

$$\mu_{G_0} = \text{plim} \frac{Ø' \hat{M}_x \hat{A}_z \hat{X} \beta_0 \beta'_0 \hat{X} \hat{A}_z \hat{M}_x ø / n}{\sigma^2 (\beta'_0 \hat{X}' \hat{M}_z \hat{M}_x \hat{M}_z \hat{X} \beta_0 / n)} \quad (4.54)$$

Turning now to $\widetilde{J}_0^m$, substituting H_T for y in (4.48), and proceeding along the lines of above calculations indicates that $\widetilde{J}_0^m$ will be non-central X^2_1 asymptotically with non-centrality parameter defined as follows:

$$\mu^2_{\widetilde{J}_0^m} = \text{plim} \frac{\left(ø' \widetilde{M}_{X_{J_0}} Z(\hat{Z}'\hat{Z})^{-1} \hat{Z}' X \beta_0 \beta'_0 X' \hat{Z} (\hat{Z}'\hat{Z})^{-1} Z' \widetilde{M}_{X_{J_0}} ø / n \right)}{\sigma^2 \left(\beta'_0 X' \hat{Z} (\hat{Z}'\hat{Z})^{-1} Z' [I_n - X(X' A_{w_{J_0}} X)^{-1} X] Z (\hat{Z}'\hat{Z})^{-1} \hat{Z}' X \beta_0 / n \right)} \quad (4.55)$$

where $\widetilde{M}_{X_{J_0}} = I_n - X (X' A_{w_{J_0}} X)^{-1} X' A_{w_{J_0}}$

The encompassing test $\widetilde{\eta}^m_{w_0}$ in (4.50) will also be asymptotically non-central X^2 under H_T but its degrees of freedom equals the number of distinct regressors in Z. Letting all k_1 regressors be distinct, for simplicity of notation, this non-centrality parameter is given by:

$$\mu^2_{\widetilde{\eta}^m_{w_0}} = \text{plim} \frac{\phi' \hat{M}_x \hat{Z} (\hat{Z}' \hat{M}_x \hat{Z})^{-1} \hat{Z}' \hat{M}_x \phi}{\sigma^2} \quad (4.56)$$

According to Das Gupta and Perlman (1974), the power of a non-central X^2 statistic increases with the value of the non-centrality parameter whereas it decreases with the degrees of freedom. Asymptotic power of the one-degree-of-freedom tests, G_0 and $\widetilde{J}_0^m$, therefore, depends on the values of μ^2_{G0} and $\mu^2_{\widetilde{J}_0^m}$. In general, from equations (4.54) and (4.55), it is not analytically possible to rank μ^2_{G0} and $\mu^2_{\widetilde{J}_0^m}$. Using Lemma 5.3 of Kullback (1968, pg. 57), it can be easily shown that $\mu^{2\ m}_{\widetilde{\eta}\, w_0} \geq \mu^2_{G0}$ and the encompassing test has more than one degree of freedom if the distinct regressors in Z exceed one. Thus, in general, the encompassing test cannot be ranked better (in terms of fixed power) than the one-degree-of-freedom tests. Q.E.D.

In general, the above power comparison remains inconclusive. However, under the specific assumption that W is the same IV matrix for H_0, H_1, and artificial regression model, H_{J_0} and H_{JA_0}, Dastoor and McAleer (1985) show that the non-centrality parameter for the J- and JA-test equals μ^2_{G0} , implying equality of powers of these tests. Asymptotic evaluation of equations (4.46), (4.47), and (4.49) along the lines of Theorem 4.10 respectively implies that $\widetilde{J}_0$, $J\widetilde{A}_0$, and $J\widetilde{A}_0^m$ are non-central statistics under H_T. The non-centrality parameters of $\widetilde{J}_0$ and $J\widetilde{A}_0$ are equal to μ^2_{G0} and the non-centrality parameter of $J\widetilde{A}_0^m$ equals to $\mu^2_{\widetilde{J}_0^m}$. Thus, the fixed power of $\widetilde{J}_0$ and $J\widetilde{A}_0$ generally equals with the power of G_0 while $\widetilde{J}_0^m$ and $J\widetilde{A}_0^m$ have the same asymptotic powers under H_T.

The Local Alternative Approach

By substituting $Z = XB + n^{-1/2}\ \Delta + \phi$ in H_1 , the local alternative of Pesaran (1982) is defined as:

$$H_L: y = XB\gamma + n^{-1/2}\ \Delta\gamma + \phi\gamma + u_1,$$

where B and are respectively $k_0\ xk_1$ and nxk_1 non-zero matrices of constants and ϕ represents terms of $o(n^{-1/2})$. To keep the notation manageable without sacrificing generality, the Z and X matrices are assumed to have no common variable; otherwise, Z, B, Δ , γ can be partitioned accordingly and H_L, can be defined in terms of distinct variables only.

THEOREM 4.11

If the number of distinct regressors in Z exceeds one, i.e., $k_1 > 1$, then G_0 has higher local power than $\tilde{\eta}^m_{w_0}$. However, the local power of $\tilde{J}^m_0$ or $J\tilde{A}^m_0$ cannot in general be ranked against the local powers of the G_0 - and encompassing tests.

PROOF

To compute the asymptotic powers, each of the test statistics needs to be asymptotically evaluated under H_L. A simple comparison of H_L with H_T and dropping the terms of o(n $^{-1/2}$) implies the following identities: $B\gamma = \beta_0$, n $^{-1/2}$ $\Delta\gamma = \phi$ and u_1 = u. Substituting these identities and Z = XB + n $^{-1/2}$ Δ in equations (4.54) through (4.56) directly yields

The following expressions for various non-centrality parameters:

$$\mu^2_{G_0} = \text{plim} \frac{\gamma'\hat{\Delta}\hat{M}_x\hat{\Delta}\gamma/n}{\sigma_1^2} \tag{4.57}$$

$$\mu^2_{\tilde{J}^m_0} = \text{plim} \frac{\gamma'\Delta\tilde{M}_{x_J}XB\gamma\gamma'B'X'\tilde{M}_{x_J}\Delta\gamma/n}{\sigma_1^2\gamma'B'X'[I_n - X(X'A_{w_J}X)^{-1}X']XB\gamma/n} \tag{4.58}$$

$$\mu^2_{\tilde{\eta}^m_{w_0}} = \text{plim} \frac{\gamma'\hat{\Delta}\hat{M}_x\hat{\Delta}\gamma/n}{\sigma_1^2} \tag{4.59}$$

where $\hat{\Delta} = A_w\Delta$, $A_{wJ_0} = W_{J_0}(W_{J_0}{}'W_{J_0})^{-1}W_{J_0}{}'$, and $W_{J_0} = [W \mid \tilde{y}_1]$ for the J-test. The non-centrality parameter for the JA-test can be deduced from (4.58) when $W_{JA_0} = [W\,\tilde{\tilde{y}}_{10}]$.

Now $\mu^2_{G0} = \mu^{2\ m}_{\tilde{\eta}\,w_0}$ and it is shown in Theorem 4.10 that, G_0 is whereas $\tilde{\eta}^m_{w_0}$ is $\chi^2_{K_1}$. Thus, it follows from Das Cupta and Perlman (1974) that G_0 has higher local power than the encompassing test when $k_1 > 1$. Again, analytical ranking of μ^2_{G0} and $\mu^2_{\tilde{J}^m_0}$ is not possible in general. However, for the special case of W_{J_0} = W_{JA_0} = W, the asymptotic-power of G_0 will equal that of $\tilde{J}^m_0$ and $J\tilde{A}^m_0$ under such a local alternative. Q.E.D.

This study has thus far discussed the asymptotic analysis of various modified non-nested tests for a presumed IV matrix. The next section is added to provide a specific IV selection criterion for the non-nested EVM.

4.6. Choice of IV for Non-Nested EVM

The standard IV selection procedures for single equation regression models, proposed in econometric textbooks, include the grouping methods by Wald (1940) and Bartlett (1949) and the ranking method by Durbin (1954). For serially uncorrected time series models, lagged values of mismeasured regressors are generally considered to be suitable instruments. In addition, there may be several other possibilities involving more data collection or mathematical ingenuity. The following discussion presents a simple and convenient method for choosing an IV within the two mismeasured non-nested models.

The primary motivation for selecting the IV matrices within the two competing non-nested models stems from the very specification of non-nested models (section 4.1): plim $(X'Z / n)$ is a non-null and non-singular matrix and, for fixed regressors, plim $(X'u_1 / n)$ and plim $(Z'u_0 / n)$ are null vectors. For mismeasured regressors, however, the latter condition may be fulfilled by assuming plim $(V'_0 V_1 / n)$ is a null matrix, where V_0 and V_1 represent the matrices of measurement errors corresponding to X and Z respectively. In addition, when the number of distinct regressors in one model, say H_0, is greater than or equal to the number of mismeasured regressors in H_1, the existence of a suitable IV can be determined.

To formalize this proposition, the following partitioning of matrices X and Z are considered.

PARTITION I Common Vs. Distinct Regressors

Let $X = [X_1 \mid \overline{X}]$ and $Z = [X_1 \mid \overline{Z}]$ where X_1 is an nxr ($r < k_0$, $r < k_1$) matrix of common regressors, while and respectively represent $k_0 - r$ and $k_1 - r$ columns of distinct regressors in H_0 and H_1 .

PARTITION II Fixed Vs. Mismeasured Regressors

Let $X = [X_f \mid X_m]$ and $Z = [Z_f \mid Z_m]$ where X_f and Z_f respectively consist of $k_0 - m_0$ and $k_1 - m_1$ columns of fixed regressors, while the m_0 and m_1

columns of mismeasured regressors in H_0 and H_1 are respectively denoted by X_m and Z_m .

When m_0 k_1 - r and m_1 k_0 - r, IV matrices for H_0 and H_1 may be chosen as $W_0 = [X_f \mid \overline{Z}]$ and $W_1 = [Z_f \mid \overline{X}]$. ($\overline{X}$) may include all distinct regressors or only a subset of the m_0 (m_1) regressors. In practice, the selection should be motivated by a consideration of those variables possessing strong asymptotic correlation with X_m (Z_m), a priori.

Using $W_0 = [X_f \mid \overline{Z}]$ and $W_1 = [Z_f \mid \overline{X}]$ as IV matrices for $X = [X_f \mid X_m]$ and $Z = [Z_f \mid Z_m]$ respectively and employing the partition inversion formulae, the following expressions can easily be derived.

$$A_{w_0} = W_0 (W'_0 W_o)^{-1} W'_0 = Ax_f + (\overline{Z}' M_{X_f} \overline{Z})^{-1} \overline{Z}' M_{X_f} \quad (4.60)$$

$$A_{w_1} = W_1 (W'_1 W_1)^{-1} W'_1 = Az_f + M_{Z_f} \overline{X} (\overline{X}' M_{Z_f} \overline{X})^{-1} \overline{X}' M_{Z_f} \quad (4.61)$$

$$\hat{X} = A_{w_0} X = [X_f \mid \hat{X}_m] \quad (4.62)$$

$$\hat{Z} = A_{w_1} Z = [Z_f \mid \hat{Z}_m] \quad (4.63)$$

$$\tilde{y}_1 = \tilde{A}_z y \quad (4.64)$$

$$\tilde{A}_x = Ax_f + M_{X_f} X_m (\hat{X}'_m M_{X_f} \hat{X}_m)^{-1} \hat{X}'_m M_{X_f} \quad (4.65)$$

$$\tilde{A}_z = Az_f + M_{Z_f} Z_m (\hat{Z}'_m M_{Z_f} \hat{X}_m)^{-1} \hat{Z}'_m M_{Z_f} \quad (4.66)$$

Here $Ax_f = X_f (X'_f X_f)^{-1} X'_f$, $Az_f = Z_f (Z'_f Z_f)^{-1} Z'_f$, $M_{X_f} = I_n - Ax_f$, $M_{Z_f} = I_n - Az_f$, $\hat{X}_m = A_{w_0} X_m$, and $\hat{Z}_m = A_{w_1} Z_m$.

Substituting the above expressions in equations (4.46), (4.47) and (4.51) through (4.53) yields the redefinitions for $\tilde{J}_0$, and G_0. Similarly, to obtain IV modified tests, $\tilde{J}_0^m$ and $J\tilde{A}_0^m$, the IV matrices for artificial regression models, H_{J_0} and H_{JA_0}, are now defined as $W_{J_0} = [W_0 \mid \tilde{y}_1]$ and $W_{JA_0} = [W_0 \mid \tilde{\tilde{y}}_{10}]$. Thus, $\tilde{J}_0^m$ and $J\tilde{A}_0^m$ are directly obtained from equations (4.48) and (4.49) by invoking the above definitions of $\widetilde{y}_1$ and $\widetilde{\widetilde{y}}_{10}$, along with, the following expressions for Aw_{J_0} and Aw_{JA_0}:

$$Aw_{J_0} = Aw_0 + Mw_0 \ \widetilde{y}_1 (\widetilde{y}'_1 Mw_0 \ \widetilde{y}_1)^{-1} Mw_0 \, , \tag{4.67}$$

$$Aw_{JA_0} = Aw_0 + Mw_0 \ \widetilde{\widetilde{y}}_{10} (\widetilde{\widetilde{y}}'_{10} Mw_0 \ \widetilde{\widetilde{y}}_{10})^{-1} \widetilde{\widetilde{y}}'_{10} Mw_0 \, , \tag{4.68}$$

where $Mw_0 = I_n - Aw_0 = Mx_f - Mx_f \ \overline{Z} (\overline{Z}' Mx_f \ \overline{Z})^{-1} \overline{Z}' \ Mx_f$.

The introduction of these IV matrices will not, in most cases, alter the asymptotic behavior of the modified non-nested tests and the earlier proofs may be readily extended to establish asymptotic correctness of size, normality with zero mean under H_0, and the consistency property of these tests. The following theorem, however, provides an exception.

THEOREM 4.12

The modified encompassing statistic, $\widetilde{\eta}^m_{w_0}$, involving IV matrices $W_0 = [X_f \mid \overline{Z}]$ and $W_1 = [Z_f \mid \overline{X}]$ is an asymptotically non-central $\chi^2_{k_1}$-r test of H_0 when in $W_1 = [Z_f \mid \overline{X}]$ consists of at least one mismeasured column.

PROOF

For an arbitrary IV matrix (W), in Theorem 4.8, $\widetilde{\eta}^m_{w_0}$ is shown to be an asymptotically central $\chi^2_{K_1}$ - r statistic under H_0. Replacing W with W_0 and W_1 and following the same procedure yields:

$$\sqrt{n}\,(\widetilde{\gamma}_1 - \widetilde{\widetilde{\gamma}}_{10}) = (\overline{Z}' Aw_1 \ \overline{Z})^{-1} \overline{Z}' Aw_1 \ [I_n - \widetilde{A}_x]\, y, \tag{4.69}$$

where Aw_1 and $\widetilde{A}_x$ are defined in equations (4.61) and (4.66). Substituting X β + u_0 for y and retaining terms of the highest order in probability yield the following:

$$\sqrt{n}\,(\tilde{\gamma}_1 - \tilde{\tilde{\gamma}}_{10}) = \sqrt{n}\ (\bar{Z}' A_{w_1}\ \bar{Z})^{-1}\bar{Z}' A_{w_1}\ u_0 - \sqrt{n}\,(\bar{Z} A_{w_1}\ \bar{Z})^{-1}\bar{Z}' A_{w_1}\ \tilde{A}_x u_0. \quad (4.70)$$

It can be easily established that plim ($\bar{Z}' A_{w_1}\ u_0 / \sqrt{n}$) is a non-zero vector when $\bar{X}$ in W_1 consists of a mismeasured column.

Thus, $\sqrt{n}\,(\tilde{\gamma}_1 - \tilde{\tilde{\gamma}}_{10})$ under H_0 is an asymptotically normal, non-zero mean vector of k_1- r variates and $\tilde{\eta}^{m}_{w_0}$ = [n $(\tilde{\gamma}_1 - \tilde{\tilde{\gamma}}_{10})'$ $(\tilde{\gamma}_1 - \tilde{\tilde{\gamma}}_{10})$ / $\tilde{V}($ $\tilde{\gamma}_1 - \tilde{\tilde{\gamma}}_{10})$] is asymptotically non-central $\chi^2_{k_1}$ - r when W_1 has a mismeasured regressor.

The asymptotic value of the non-centrality parameter may be given by:

$$\mu^2_{\tilde{\eta}\ w_0^m} = \text{plim}\ \frac{u'_0 A_{w_1}\ \bar{Z}[\bar{Z} A_{w_1}\ \tilde{M}_X \tilde{M}'_X A_{w_1}\ \bar{Z}]^{-1}\ \bar{Z} A_{w_1}\ u_0}{\sigma_0^2} \quad (4.72)$$

COROLLARY 4.12

The non-centrality parameter in equation (4.72) is generally unidentified.

PROOF

Invoking the EVM assumption, plim (X' u_0 / n) = - β Q_{V_0} , and assuming, for simplicity of notation, W_1 = X implies that

$$\mu^2_{\tilde{\eta}\ w_0^m} = \text{plim}\ \frac{n\,\beta' Q_{V_0} Q_{XX}^{-1}\ Q_{X\bar{Z}}\,[\ \text{plim}\ \bar{Z} A_{w_1}\ \tilde{M}_X \tilde{M}'_X A_{w_1}\ \bar{Z}/n]^{-1}\ Q_{\bar{Z}X}\,Q_{XX}^{-1} Q_{V_0}\,\beta}{\sigma_0^2}$$

The variance-covariance matrix for measurement errors in X, Q_{V_0} , is generally unidentified and so will be $\tilde{\mu}^{m}_{w_0}$. Q.E.D.

Similarly, the above results may readily be applied to Dastoor's R-procedure, Dastoor and McAleer's (1985) LM version of the R-procedure, and conventional F-test of a compound model. Any combination of W_0 and W_1 to obtain W_c - an IV matrix for compound regressor matrix $X_c = [X \mid \overline{Z}]$ - provides an equality between X_c and W_c . Thus, inconsistent estimators for the compound model will lead to an asymptotic non-central F-test.

The next theorem considers the special case when all the regressors of one model act as the IV for other model and vice versa, i.e., $W_0 = Z$ and $W_1 = X$. In practice, such situations may arise when all fixed regressors in H_0 and H_1 (if any) are common and all distinct regressors in both models are mismeasured, i.e., $X_f = Z_f = X_1$, $= X_m$ and $\overline{Z} = Z_m$. If $m_0 = m_1$ ($k_0 = k_1$) and the measurement errors of X_m and Z_m are assumed asymptotically uncorrelated, then it may appear appropriate at first glance to consider Z and X as IV matrices for X and Z.

THEOREM 4.13

The modified test statistics $\widetilde{J}_0$, $J\widetilde{A}_0$, $\widetilde{J}_0^m$, $J\widetilde{A}_0^m$ and G_0 equal zero if $W_0 = Z$ and $W_1 = X$.

PROOF

Substituting $W_0 = Z$ and $W_1 = X$ in equations (4.60) through (4.67) and noting that X'Z and Z'X are square and non-singular matrices, then $\widetilde{y}_1 = Z(X'Z)^{-1}X'y$, $\widetilde{\widetilde{y}}_{10} = Z(X'Z)^{-1}X'X(Z'X)^{-1} Z'y$ and $\widetilde{M}_X = I_n - X(Z'X)^{-1} Z'$ implies that $\widetilde{J}_0$ and $J\widetilde{A}_0$ in equations (4.46) and (4.47) will not exist since

$$y'_1 \widetilde{M}_X y = \widetilde{\widetilde{y}}'_{10} \widetilde{M}_X y = 0.$$

The non-existence of and follows from the non-existence of Aw_{J_0} and Aw_{JA_0} in equations (4.68) and (4.69) since $\widetilde{y}'_1 Mw_0 \widetilde{y}_1 = \widetilde{\widetilde{y}}'_{10} Mw_0 \widetilde{\widetilde{y}}_{10} = 0$ when $Mw_0 = I_n - Z(Z'Z)^{-1} Z'$. Similarly, the test statistic G_0 in equation (4.51) collapses since $\widetilde{u}'_1 A_w \widetilde{u}_1 = 0$ when $A_w = A_z$. Q.E.D.

However, to avoid these problems, the earlier conditions m_0 k_1 - r and m_1 k_0 - r may further be tightened by establishing strict inequalities, i.e., $m_0 < k_1$ -

r and $m_1 < k_0$ - r, and including only a subset of $\overline{Z}$ and $\overline{X}$ in W_0 and W_1 respectively.The following theorem involves a special case where $W_0 = [X_f \mid Z_f] = W_1$ for $X = [X_f \mid X_m]$ and $Z = [Z_m \mid Z_f]$. Such a case may arise if X_f and Z_f consist of only distinct columns and the number of fixed regressors in each model at least equals the number of mismeasured regressors in other model, i.e., $m_0\ k_1 - m_1$ and $m_1\ k_0 - m_0$.

THEOREM 4.14

If $W_0 = [X_f \mid Z_f] = W_1$ then $\tilde{\eta}^m_{w_0}$ and G_0 do not exist while $\tilde{J}_0 = J\tilde{A}_0$ and $\tilde{J}^m_0 = J\tilde{A}^m_0$ exist as appropriate tests of H_0.

PROOF

Applying the partition inversion rule to $W = [X_f \mid Z_f] = W_1$, it can be shown that

$$
\begin{aligned}
A_{w_0} &= A_{x_f} + M_{x_f} Z_f (Z'_f M_{x_f} Z_f)^{-1} Z'_f M_{x_f} \\
&= A_{z_f} + M_{Z_f} X_f (X'_f M_{Z_f} X_f)^{-1} X'_f M_{Z_f} \\
&= A_{w_1} \qquad (4.74)
\end{aligned}
$$

substituting $A_{w_0} = A_{w_1}$ in equations (4.66) and (4.67) indicates the following:

$$\overline{Z}' A_{w_1} \tilde{A}_x = Z' A_{w_1}\ , \qquad (4.75)$$

$$\tilde{A}_z \tilde{A}_x = \tilde{A}_z\ , \qquad (4.76)$$

$$\tilde{A}_x \tilde{A}_z = \tilde{A}_x\ . \qquad (4.77)$$

$$\tilde{A}'_z \neq \tilde{A}_z\ , \qquad (4.78)$$

$$\tilde{A}'_x \tilde{A}_z \neq \tilde{A}'_x \neq \tilde{A}_x \qquad (4.79)$$

Substituting the equation (4.75) in equation (4.70) implies that $\sqrt{n}\,(\tilde{\gamma}_1 - \tilde{\tilde{\gamma}}_{10}) = 0$. Similarly, the expression (4.76) and (4.77) cause the numerator of G_0 in equation (4.52) to collapse to zero, i.e., $[y'(\hat{A}_x \hat{A}_z \hat{A}_x - \hat{A}_z) y / n] = 0$. Thus, $\tilde{\eta}^m_{w_0}$ and G_0 do not exist.

The inequalities (4.78) and (4.79), on the other hand, readily ensure the existence of $\widetilde{J}_0$ and $\widetilde{J}_0^m$ since $\sqrt{n}\,(\widetilde{y}'_1\, \widetilde{M}_X y)$ and $\sqrt{n}\,(\,\widetilde{y}'_1\,[\,I_n - X(X'\,A_{W_{J_0}}\,X)^{-1}\,X'\,A_{W_{J_0}}]y)$ in equations (4.46) and (4.48) remain well denned. Moreover, $\widetilde{A}_z\,\widetilde{A}_x = \widetilde{A}_z$ establishes the equality between $\widetilde{y}_1$ and $\widetilde{\widetilde{y}}_{10}$. Hence, $\widetilde{J}_0 = J\widetilde{A}_0$ and $\widetilde{J}_0^m = J\widetilde{A}_0^m$ will exist as suitable tests of H_0. Q.E.D.

The above proof for $\widetilde{\eta}_{w_0}^m$ can be readily extended to the conventional F-test, Dastoor's R-test and LM type extension of the R-test, so that these compound model approaches similarly fail to be appropriate tests when $W_0 = [X_f \mid Z_f] = W_1$.

Finally, the equations (4.54) through (4.59) — defining the non-centrality parameters of various tests under the 'fixed' and 'local' alternatives - may in general be redefined by substituting appropriate expressions from (4.60) through (4.69). The results in theorems 4.10 and 4.11, implying no general ranking of the fixed and local asymptotic powers of various modified test, will typically remain unchanged.

4.7. Conclusion

This chapter provides a significant contribution to the existing literature on non-nested hypothesis testing. By imposing the EVM structure on one or both non-nested models, it is shown that the conventional non-nested tests asymptotically converge to non-central limits under the null (H_0). Although the tests are consistent in general, there exists a non-conformity between the true and nominal sizes under H_0. The asymptotic behavior of non-nested tests remains unaltered if the regressors in non-nested models are fixed or non-stochastic while only the regress and involves measurement errors.

The inconsistent least squares estimators for mismeasured non-nested model(s) are replaced with the consistent IV estimators and conventional non-nested tests are modified accordingly. A simple replacement of the least squares estimators with their IV counterparts fails to correct the size problem for the linearized Cox and Mizon's (1984) variance encompassing tests. On the other hand, such substitutions provide IV-based J- and JA-tests ($\widetilde{J}_0$ and $J\widetilde{A}_0$) which are asymptotically normal with zero mean under H_0 and are asymptotically unbounded under H_1. Unfortunately, these tests cannot be interpreted and computed as the t-ratios from augmented regression models which destroy the computational feasibility of the linear embedding approach. Therefore, two more modified statistics $\widetilde{J}_0^m$, and $J\widetilde{A}_0^m$ are, formally obtained as t-ratios from the estimation of artificial regression models. These tests follow an asymptotically standard normal distribution under H_0 and also satisfy the consistency property.

Another IV- modified test based on slope parametric encompassing principle, which may be also interpreted as Dastoor's (1983) R-test or conventional F-test involving IV estimation of a comprehensive model, is developed as an asymptotically $\chi^2_{k_1}$ - r statistic under H_0. To prove consistency, the statistic is shown to be asymptotically unbounded when H_1 is true. However, it is argued that an IV-modified expression for the variance encompassing test will be unidentified.

The fixed and local approaches are used to compare asymptotic powers of various modified tests. It is shown that, if different instruments are used for H_0 and H_1, then no ranking of power can be determined in general. However, the G-test possesses higher local power asymptotically than the modified encompassing test if the number of distinct regressors in the alternative model exceeds one.

The selection of IV matrices for EVM non-nested models has been given careful consideration in this paper. If both non-nested models are specified as EVM, then, under certain conditions, the IV matrices may be chosen from the regressors available within the two models. To fulfill these requirements, the number of distinct regressors in one model must exceed or equal the number of mismeasured regressors in the other model; a mismeasured distinct regressor may act as IV for the other model if its measurement errors are asymptotically uncorrelated with measurement errors in other model. However, the existence of certain modified test statistics further requires that the IV matrix for one model does not include all regressors in other model. Moreover, the modified Wald encompassing, the conventional F-, and the R-test are shown to be asymptotically non-central tests of H_0, and the non-centrality parameters are generally unidentified if a mismeasured distinct repressor in one model act as IV for the other model. Another special choice of the IV matrix, as the fixed repressors in H_0 and H_1 leads to non-existence of the G-test, F-test, and modified forms of the Wald encompassing test, R-test and LM type R-test. Howver, $\widetilde{J}_0 = J\widetilde{A}_0$ and $\widetilde{J}_0^m = J\widetilde{A}_0^m$ are well defined tests of H_0 in this instance.

This chapter completes the analytical evidence on the asymptotic behavior of various modified non-nested tests. In the next chapter, Monte Carlo experiments substantiate these asymptotic claims and evaluate the finite sample behavior of these tests.

Chapter 5

NON-NESTED EVM TESTS: SOME MONTE CARLO EVIDENCE

5.1. INTRODUCTION

Theoretical properties of non-nested tests in general (and the modified non-nested tests for errors-in-variables models in particular) are limited to asymptotic results, whereas real world econometric analysis deals with finite samples. Earlier studies by Pesaran (1974, 1982a), Davidson and MacKinnon (1982) and Godfrey and Pesaran (1983) provide some evidence on the small sample behavior of various non-nested tests. Their design specificity however, limits the scope of their findings to non-nested linear regression models when all regressors are non-stochastic or exogenous. Monte Carlo evidence on finite sample performance of the modified non-nested tests of mismeasured regression models is provided in this chapter.

The first section in this chapter describes the design of the Monte Carlo experiments. The second section details computational aspects. The results of various experiments are presented in the third section, and the last section provides a summary of the major findings.

5.2. THE DESIGN OF EXPERIMENTS

Non-nested errors-in-variables models may be written as:

$$H_0: y = X\beta + u_0, \tag{5.1}$$
$$H_1: y = Z\gamma + u_1, \tag{5.2}$$

where the nxk_1 matrices of explanatory variables, X and Z, contain nonoverlapping variables and consist of some or all mismeasured variables; i.e., $X = X_T + V_0$ and $Z = Z_T + V_1$. X_T and Z_T represent the correctly measured components of X and Z whereas V_0 and V_1 are the matrices of measurement errors.

The number of mismeasured regressors in X and Z are denoted by m_0 and m_1, respectively, so that, m_0 and m_1 must also be the number of non-zero columns in V_0 and V_1.

Each non-zero column in V_0 and V_1, denoted v_{0i} and v_{1j} for $i = 1,...,m_0$ and $j = 1,...,m_1$, is assumed to be $N(0, \sigma^2_{voi} I_n)$ and $N(o, \sigma^2_{v1j} I_n)$ where σ^2_{voi} and σ^2_{v1j} are constants.

The Monte Carlo experiments aimed at performance evaluation of the IV-based or modified tests for H_0 and H_1 above are described in the following four subsections.

The first subsection defines the 'true' data generating process (DGP) and its 'fixed' alternative followed by a discussion of the mismeasured models.

Since these EVM specifications are to be estimated with instrumental variables, the third section discusses alternative choices for these IV matrices and the fourth subsection deals with parameter selection.

5.2.1. Description of the True' DGP and 'Fixed' Alternative

The experimental design described below is a reproduction of the models in Pesaran (1982a) and Godfrey and Pesaran (1983), with the exception of the introduction of mismeasured regressors.

By replicating their Monte Carlo design, a direct comparison may be made between their non-stochastic regressor models and the EVM tests studied here. The true, albeit unobservable, DGP for each replication of the Monte Carlo experiment is specified by the following model:

$$H_T: y = X_T\beta + u_T \tag{5.3}$$

The n observations for each of the k_0 columns in X_T are generated as iid standard normal variates, while β is chosen as a unit vector of order k_0x1.

The error term in (5.3) is generated as $u_T \sim N(0, \sigma^2_T I_n)$, whose variance, σ^2_T, is given by

$$\sigma_T^2 = \frac{\beta'\beta(1 - R^2)}{R^2} = \frac{k_0(1 - R^2)}{R^2} \tag{5.4}$$

where R^2 is the coefficient of multiple determination for H_T.

The 'false' non-nested alternative model is given by:

$$H_F: y = Z_T \gamma + u_F \tag{5.5}$$

where y represents the true DGP in (5.3), Z_T is an nxk$_1$ matrix of explanatory variables, and u_F represents an nxl vector of unexplained discrepancies between y and Z_{Tr}. The condition of non-orthogonality between X_T and Z_T is ensured by generating the ith column in Z_T as follows:

$$Z_{Ti} = \lambda i X_{Ti} + e_{Ti} \quad (i = 1,\ldots,k_I) \tag{5.6}$$

where $e_{Ti} \sim N(0, I_n)$, $\lambda i = p_i / \sqrt{1 - p_i^2}$ and p_i is the simple correlation coefficient between X_{Ti} and Z_{Ti}. It is assumed for simplicity that p_i = p, a constant for all i, and that X_T and Z_T contain no common regressors; that is, $p \neq 1$ for any i and p_i is non-zero for all i. Assuming that the value of p is fixed as n increases, H_F may be viewed as the 'fixed ' alternative of Davidson and MacKinnon (1982).

Although Godfrey and Pesaran (1983) consider both equal and unequal numbers of regressors, the case of $k_0 = k_I$ = k will be undertaken here.

5.2.2. Description of the EVM

The mismeasured non-nested models, H_0 and H_1, are obtained by transforming the correctly measured regressor matrices X_T and Z_T into the mismeasured regressor matrices X and Z. The m_0 independent columns of measurement errors for V_0 were generated as $v_{oi} \sim N(0, \sigma^2_{voi} I_n)$ for i = 1,...,m_0: similarly, $v_{1j} \sim N(0, \sigma^2_{v1j} I_n)$ for j = 1,...,m_1. By adding V_0 to the first m_0 columns of X_T and V_1 to the last m_1 columns of Z_T, the mismeasured regressors are defined as $X = [X_m \mid X_f]$ and $Z = [Z_f \mid Z_m]$ where X_m and Z_m denote mismeasured columns in X and Z while X_f and Z_f represent the correctly measured regressors. The number of mismeasured regressors and the variances of the measurement errors are expected to influence the performance of the modified non-nested tests. However, to keep the number of experiments within limits, it is assumed that $m_0 = m_1 = m$ and $\sigma^2_{voi} = \sigma^2_{v1j} = \sigma^2_m$ for all i and j.

5.2.3. Description of the IV

As discussed in section 4.5, appropriate instrumental variables for the non-nested tests of errors-in-variables models may, under certain conditions, be chosen as regressors of one model acting as instrumental variables for the other model. In the experiment detailed below, the null and alternative hypotheses will each contain the same number of correctly measured and mismeasured regressors. (For example, when $k = 2$, $m_0 = m_1 = 1$.) In this instance, it is appropriate to choose the correctly measured regressor(s) in H_0 (H_1) as the instrument(s) for the mismeasured regressor(s) in H_0 (H_1). More formally, the IV matrix for these experiments is given by $W_0 = [Z_f \mid X_f] = W_1$. Thus, $W_0 = W_1$ is obtained by combining the correctly measured columns in X and Z.

5.2.4. Parameter Selection

The IV selection from the regressors of competing non-nested models provides a novel and specific approach to the general problem of instrumental variable specification. To fully examine the properties of the EVM non-nested tests, the chosen experimental design involves the 320 design points defined by the following chosen parametric values.

$(k, m) = (2, 1), (4, 2)$

$R^2 = (0.99, 0.95, 0.80, 0.50, 0.30)$

$\rho^2 = (0.30, 0.50, 0.80, 0.95)$

$\sigma^2_m = (1.0, 0.25)$

$n = (20, 40, 60, 100)$

It is expected that, *ceteris paribus,* the performance of the non-nested EVM tests will worsen with decreases in the sample size (n) and R^2, and with increases in ρ^2 and σ_m^2. For instance, since σ_m^2 is the mismeasurement error variance, as σ_m^2 approaches zero, this experiment collapses to that of Godfrey and Pesaran (1983) for non-stochastic regressors. On the other hand, a poor fit of the true DGP, as measured by a low value for R^2, will make it more difficult for the tests to identify the 'true' model.

5.3. The Computational Details

In the past most of the computational works done using FORTRAN 77 and compiled on the AMDAHL 5890. Now a days R language is being used to perform all necessary computations of the modified tests and performance indicators. Some of the selected computations are detailed below.

The generation of the null and alternative hypotheses requires several independently and identically distributed normal deviates; the matrices X_T, u_T, e_T, v_{oi}, and V_{1j} were generated by invoking the DRNNOA subroutine of the IMSL/STAT Library version 1.0. For each of the 320 experiments defined in section 5.1.4, 500 replications were performed. By setting the number of replications at 500, there is a 95% probability that the estimated and nominal Type-I errors will differ by no more than 0.02: on this point, see Pesaran (1974, pg. 160).

For the chosen instrument $W_0 = W_1 = [Z_f \mid X_f]$, the non-nested test statistics G_0, $\tilde{\eta}^m_{w0}$ and $\tilde{\eta}_{LM0}$ do not exist: see theorem 4.14. Also in this instance, $\tilde{J}_0 = J\tilde{A}_0$ and $\tilde{J}_0^m = J\tilde{A}_0^m$ Computationally, it is easier to work with the squares of these test statistics, so that the statistics computed for these experiments will be $\tilde{J}_0^2$ and $\tilde{J}_0^{m^2}$ each following a X_1^2 distribution under H_0.

By interchanging the roles of H_0 and H_1, the values $\tilde{J}_1^2$ and $\tilde{J}_1^{m^2}$ serve as the test of H_1. The performance indicators calculated for each experiment are the mean, variance, skewness and kurtosis of the empirical distribution, the x^2 goodness-of-fit test, and Type-I error, Type-II error, and power measures for $\tilde{J}_0^2$ and $\tilde{J}_0^{m^2}$.

The mean, variance and coefficients of skewness and kurtosis for the sampling distributions of $\tilde{J}_0^2$ and $\tilde{J}_0^{m^2}$ are calculated by invoking the DUVSTA subroutine of the IMSL/ STAT Library. Similarly, by calling DCHIGF, the Chi-square goodness-of-fit measures are obtained. Since the sample observations for each statistic are divided into 20 equal groups, the goodness-of-fit test will have 19 degrees of freedom.

The Type-I errors, Type-II errors, and powers of these tests are tabulated at the three conventional significance levels: $\alpha = 0.01, 0.05, 0.10$. In order to determine appropriate critical values for each significance level, the IMSL/STAT subroutine DCHIIN was invoked. The empirical Type-I and Type-II errors for a test, say $\tilde{J}^2$, are respectively defined as the proportion of times in 500 replications that $\tilde{J}_0^2$ is greater than, and $\tilde{J}_1^2$ is less than, the critical value of X_1^2 for a certain value of α. On the other hand, the power of a test represents the proportion of times that neither a Type-I nor Type-II error occurs. The standard errors for the Type-I error, Type-II error and power of each test are also estimated. Following Pesaran (1974, pg. 161), the estimated standard error, say for the size of a test, is given by $\sqrt{\hat{p}(1-\hat{p})/500}$ where $\hat{p}$ denotes the estimated Type-I error.

5.4. The Results of Experiments

The full sets of Monte Carlo results are reported in 80 tables which are organized as follows. Any one table presents the performance indicators for $\tilde{J}_0^2$ and $\tilde{J}_0^{m^2}$ for all values of n = 20, 40, 60, and 100, permitting a direct evaluation of the approach to the asymptotic distribution. From here, the tables are organized into four major groups: Tables 5.1, 5.2, 5.3, and 5.4. Within each group may be found all combinations of the values of R^2 = (0.99, 0.95, 0.80, 0.50, 0.30) and = (0.30, 0.50, 0.80, 0.95) for a given set of values for k, m, and σ_m^2. In particular,

Tables 5.1.1 - 5.1.20 set k = 2, m = 1, $\sigma_m^2 = 1.00$,

Tables 5.2.1 - 5.2.20 set k = 4, m = 2, $\sigma_m^2 = 1.00$,

Tables 5.3.1 - 5.3.20 set k = 2, m = 1, $\sigma_m^2 = 0.25$,

Tables 5.4.1 - 5.4.20 set k = 4, m = 2, $\sigma_m^2 = 0.25$.

Appendix A tabulates the complete set of parameter values found in each of these tables.

For different sample sizes (n) and a given set of other parametric values, each table reports the estimated mean (M), variance (V), coefficients of skewness (S) and kurtosis (K), and the goodness-of-fit measure (x^2) for $\tilde{J}_0^2$ and $\tilde{J}_0^{m^2}$. The estimated Type-I and Type-II errors and power of each statistic at 1%, 5% and 10% significance levels, with standard errors in parentheses, are reported in the last nine columns of each table.

The observed size estimates in all tables that depart significantly (i.e., differ by more than two standard errors) from their nominal values are marked with an asterisk. Similarly, the goodness-of-fit test (x^2) exceeding X^2_{19} at the *a* = 5% level is so marked ($X^2_{19,0.05}$ 30.144), indicating a lack of fit between the empirical distribution and the theoretical x^2_1 distribution. The theoretical mean of these distributions is one (the degrees of freedom of the x^2_1 distribution), so that the amount by which the empirical mean (M) differs from 1 is the estimated bias. Analogously, the theoretical variance is two (twice the degrees of freedom of the x^2_1distribution). Thus, a cursory examination of each table reveals the behavior of a test for different sample sizes as well as a comparison between $\tilde{J}^2$ and $\tilde{J}^{m^2}$ for a given set of parameters. However, evaluating the tests for different parametric values involves inter-table comparisons. Some of the interesting results of this experiment are now discussed.

As expected, the observed means, variances, measures of goodness-of-fit, and sizes of both tests generally approach their theoretical counterparts as the sample size gets larger. For example, the size of $\tilde{J}^2$ in Table 5.1.1 moves from 0.046 at n = 20 to 0.082 at n = 100 when = 0.10; similarly, the size of $\tilde{J}^{m^2}$ ranges between 0.068 and 0.074. Increasing sample sizes also yield higher powers for those tests with correct sizes; e.g., in table 5.1.3 at $\alpha = 0.10$, $\tilde{J}^{m^2}$ has the correct size for all sample values while its power increases from 27% at n = 20 to 67% at n = 100.

The relative performance of $\tilde{J}^2$ and $\tilde{J}^{m^2}$ can also be seen from these results. A careful examination of each table, in all of the four sets, reveals that the observed significance levels of $\tilde{J}^2$ are generally less than the corresponding values of $\tilde{J}^{m^2}$.

However, the means of both tests are not usually farther apart (with the exception of the results reported in tables 16 and 17 of each group, in which $R^2 = 0.80, 0.95$ and $\rho^2 = 0.95$), suggesting that the lower size of $\tilde{J}^{m^2}$ is most likely caused by the lower values of its variance and kurtosis. Moreover, the power of $\tilde{J}^{m^2}$ generally exceeds that of $\tilde{J}^2$ (when each test has the correct size). As an example, in those cases where $\tilde{J}^2$ and $\tilde{J}^{m^2}$ each have correct Type-I errors, selected powers of these tests have been reproduced below when n = 100.

Selected Estimated Powers

$R^2 = 0.99 \quad \rho^2 = 0.30$

	TABLE 5.1.1 **α = 0.01)**	**TABLE 5.2.1** **(α = 0.01)**	**TABLE 5.3.1** **(α = 0.05)**	**TABLE 5.4.1** **(α = 0.01)**
$\tilde{J}^2$ $\tilde{J}^{m^2}$	0.374 0.670	0.320 0.676	0.890 0.942	0.824 0.970

$R^2 = 0.30 \quad \rho^2 = 0.95$

	TABLE 5.1.20 **(α = 0.01)**	**TABLE 5.2.20** **(α = 0.01)**	**TABLE 5.3.20** **(α = 0.05)**	**TABLE 5.4.20** **(α = 0.05)**
$\tilde{J}^2$ $\tilde{J}^{m^2}$	0.020 0.024	0.018 0.024	0.058 0.066	0.082 0.086

The higher power of $\tilde{J}^{m^2}$ over $\tilde{J}^2$ is quite encouraging, since the modified test, $\tilde{J}^{m^2}$, is also the easiest of the two tests to compute. (Recall that $\tilde{J}^m$ can be viewed as the direct t-ratio in a compound model, whereas $\tilde{J}$ cannot be calculated on existing software). An important caveat to this claim of high power also needs to be made - in some instances, the $\widetilde{J}^{m^2}$ test fails to reach the correct size even in samples as large as 100: see, for example, Table 5.1.17.

The Monte Carlo results in Godfrey and Pesaran (1983) measure an expected decline in the power of non-nested tests *ceteris paribus* with decreasing ρ^2. The same trend applies to the EVM non-nested tests. For example, the power of $\widetilde{J}^{m^2}$ as R^2 decreases (Tables 5.1.6 through 5.1.10) when n = 100 and a = 0.10 are 0.810, 0.792, 0.746, 0.622 and 0.464. On the other hand, the effect of in this experiment is not the same as those reported by Godfrey and Pesaran for the case of non-stochastic regressor models, since ρ^2 plays two competing roles here. First, the increased value of ρ^2 increases the efficiency of the IV estimators of H_0 which may increase the power of the modified tests. Second, with increasing values of ρ^2, differentiating between the null and fixed alternative hypotheses becomes increasingly difficult and the power of a non-nested test may decline. The power of $\widetilde{J}^2$ and $\widetilde{J}^{m^2}$ (when these tests have correct sizes) ρ^2 increases in all four sets when increases from 0.30 to 0.50 while other parameters are fixed. For example, the powers of $\widetilde{J}^2$ are reproduced below for those instances in which R^2 = 0.99, n = 100 and α = 0.01. From these results, it is noted that the power of $\widetilde{J}^2$ (and also that of $\widetilde{J}^{m^2}$ not tabulated here) increases up to a point (ρ^2 = 0.50) and then begins to decline. In one instance, the drop in power is quite dramatic — note for group 4 that power drops from 94% when ρ^2 = 0.50 to 45% when ρ^2 = 0.95. Thus, it is interesting to conclude that for this particular experimental design, at lower levels of ρ^2, an increase from 0.30 to 0.50 causes the efficiency effect of the IV estimators of H_0 to dominate, causing an increase in the power of $\widetilde{J}^2$ and $\widetilde{J}^{m^2}$. On the other hand, further increases in ρ^2 (greater than 0.50) cause declining powers due to the increased closeness between the null and fixed alternative hypotheses.

Selected Estimated Powers

	k= 2 m = 1 σ_m^2 = 1.00	k= 4 m = 2 σ_m^2 = 1.00	k= 2 m = 1 σ_m^2 = 0.25	k= 4 m = 2 σ_m^2 = 0.25
ρ 2 = 0.30	0.374	0.320	0.832	0.824
ρ 2 = 0.50	0.484	0.458	0.950	0.940
ρ 2 = 0.80	0.310	0.300	0.938	0.942
ρ 2 = 0.95	0.052	0.052	0.470	0.454

The effects of (k, m) on the performance of $\tilde{J}^2$ and $\tilde{J}^{m^2}$ can also be isolated. In general, both tests exhibit the correct Type-I error more often when (k, m) = (2, 1) than when (k, m) = (4, 2). Moreover, for these cases in which valid comparisons can be made, power declines when (k, m) increases from (2, 1) to (4, 2). However, the differences in powers are usually less pronounced for higher values of n and R^2 and low values of ρ^2 , i.e., n = (60, 100) , R^2 = (0.99, 0.95, 0.80) and ρ^2 = (0.30, 0.50), whereas differences in power are otherwise quite substantial. Finally, a divergence between the estimated powers of the $\tilde{J}^2$ and $\tilde{J}^{m^2}$ tests is more likely to be found as k and m increase.

The performance indicators generally reveal marked improvements when σ_m^2 , the mismeasurement error variance, declines from 1.00 to 0.25. In Tables 5.3.1 through 5.3.20 and 5.4.1 through 5.4.20, where σ_m^2 = 0.25, the observed values of various measures are mostly in close proximity to their true values; the notable discrepancies in the mean, variance, and size of $\tilde{J}^{m^2}$ occur when n = 20, R^2 = 0.95, 0.80, and ρ^2 = 0.95 (see Tables 5.3.17, 5.3.18, 5.4.17 and 5.4.18). The effect of σ_m^2 on power is typified in the following, where the powers of $\tilde{J}^2$ for n = 20, 100, a = 1%, R^2 = 0.99, and ρ^2 = 0.30 in Tables 5.1.1 and 5.1.2 are respectively compared against those in Tables 5.3.1 and 5.4.1.

Selected Estimated Powers

	k= 2 m = 1 $\sigma_m^2 =$ 1.00	k= 4 m = 2 $\sigma_m^2 = 1.00$	k= 2 m = 1 $\sigma_m^2 = 0.25$	k= 4 m = 2 $\sigma_m^2 = 0.25$
n = 20	0.350	0.672	0.228	0.542
n = 100	0.670	0.978	0.676	0.970

Interestingly, power actually declines when σ_m^2 declines in the small sample case (n = 20). It is only when n = 100 that improvements in power are associated with declining mismeasurement errors for the $\widetilde{J}^{m^2}$ test. Finally, the size and power of $\widetilde{J}^{m^2}$ for $\sigma_m^2 = 0.25$ are compared with those of Type-I Error and Power Comparisons of $\widetilde{J}^{m}$ and J^2.

The Type-I errors and powers of J^2 (in the columns marked $\sigma_m^2 = 0$) are reproduced from McAleer (1987) as originally computed by Godfrey and Pesaran (1983). The asterisk here denotes the Type-I errors that are significantly different from $a = 0.05$.

Godfrey and Pesaran's (1983) values for J^2 where $\sigma_m^2 = 0$ (i.e., the case of non-stochastic regressors). Fixing $R^2 = 0.50$ and $a = 0.05$, provides an expected decrease in power with increasing values of ρ^2, σ_m^2 and k. There exists a marked difference between the powers of J^2 and $\widetilde{J}^{m^2}$ for lower values of ρ^2 while it is insignificant when $\rho^2 = 0.95$. It may also be expected that such differences will diminish in large samples (n = 100) or with higher values of R, since the power of $\widetilde{J}^{m}$ in these cases reaches the maximum.

However, the published results of Godfrey and Pesaran (1983) are not available for such parametric values and no direct comparisons are possible to prove the point. Finally, the small sample over-rejection of H_0 seems to be a serious problem for J^2 when k = 4 and $\rho^2 = 0.30$, 0.50, and 0.80 whereas $\widetilde{J}^{m^2}$ has correct sizes.

Type-I Error and Power Comparisons of $\widetilde{J}^{m^2}$ and J^2
$n = 20$, $R^2 = 0.50$, $a = 0.05$

	Type-I Error		Power	
	$\sigma_m^2 = 0.25$	$\sigma_m^2 = 0.00$	$\sigma_m^2 = 0.25$	$\sigma_m^2 = 0.00$
$k = 2, \rho^2 = 0.30$	0.050	0.062	0.376	0.834
$k = 2, \rho^2 = 0.50$	0.070	0.044	0.324	0.766
$k = 2, \rho^2 = 0.80$	0.072	0.054	0.200	0.418
$k = 2, \rho^2 = 0.95$	0.054	0.052	0.098	0.126
$k = 4, \rho^2 = 0.30$	0.052	0.142*	0.236	0.778
$k = 4, \rho^2 = 0.50$	0.060	0.108*	0.264	0.732
$k = 4, \rho^2 = 0.80$	0.070	0.074*	0.160	0.416
$k = 4, \rho^2 = 0.95$	0.044	0.072	0.094	0.108

In general, the EVM non-nested tests have performed as expected. For this particular Monte Carlo design, these tests appear to be most useful when R^2 and n are relatively large (i.e., $R^2 > 0.80$ and $n > 40$) and is moderately low (i.e., $\sigma_m^2 = 0.25$). The effect of on power is ambiguous.

5.5. Conclusion

This chapter, for the first time in the econometric literature, provides Monte Carlo evidence on the behavior of non-nested tests that involve IV estimation of mismeasured regression models. The design of the Monte Carlo experiment involves a true DGP and a fixed alternative hypothesis suggested by Godfrey and Pesaran (1983), while some of the explanatory variables in both models are then designed to be mismeasured. Instrumental variables are chosen to be the correctly measured regressors in the competing model. Different combinations of the values of n, R^2, k, m and σ_m^2 are specified; the mean, variance, coefficients of skewness and kurtosis, measure of goodness-of-fit, Type-I errors, Type-II errors and powers of two modified tests are then derived.

The results from the 320 outcomes are obtained for the squared values of $\tilde{J} = J\tilde{A}$ and $\tilde{J}^m = J\tilde{A}^m$. The mean, variance, size, goodness-of-fit measure, and power of $\tilde{J}$ and $\tilde{J}^m$ significantly improve as the sample size increases. For all combinations of R^2 and ρ^2, both tests have the correct size in large samples. The powers of $\tilde{J}^m$ are generally higher than those of $\tilde{J}$, *ceteris paribus*. Powers of the and $\tilde{J}^m$ tests are reasonably high for those combinations involving n = (100, 60), R^2 = 0.99, 0.95, 0.80) and ρ^2 = (0.30, 0.50). A substantial increase in the powers of $\tilde{J}$ and $\tilde{J}^m$, for these parametric values, occurs when σ_m^2 decreases from 1.0 to 0.25. In general, power is directly related to n and R^2, and inversely related to σ_m^2 and k. One interesting outcome of these experiments involves the two competing roles of ρ^2 in the determination of the power of $\tilde{J}$ and $\tilde{J}^m$. An increase in ρ^2 from 0.30 to 0.50 causes a significant increase in the power of $\tilde{J}$ and $\tilde{J}^m$, while any further increase in ρ^2 (i.e., 0.80 or 0.95) causes power to decline. The former increase in power may be attributed to increased efficiency of the IV estimators of H_0, and the latter decrease in power may be related to the increased closeness between the null and fixed alternative hypotheses.

These Monte Carlo results are necessarily limited to the chosen experimental design. Nonetheless, they establish a plausible case for testing non-nested mismeasured regression models using readily available instrumental variables. The computational convenience of $\tilde{J}^m$, supplemented with its generally high power, enhances its practical utility for applied researchers.

Chapter 6

POWER COMPARISON OF NON-NESTED TESTS

6.1. INTRODUCTION

Davidson and Mackinnon (2001) proposed a more sophisticated bootstrapped J-test (fast double bootstrapped - FDB) for non-nested linear regression models and found a significant improvement over the ordinary bootstrap. The FDB procedure is, in fact, about twice as expensive as the usual single bootstrap. Although the FDB is not as widely applicable as the double bootstrap, it can be applied to a greater range of econometric tests, including the J test and other non-nested tests. Recently, Davidson and MacKinnon (2002) developed a theoretical approach which determines the finite sample distribution of the J-test precisely. Based on the approach, a Monte Carlo simulation experiment showed that in most cases, the J-test performs very well in finite samples if it is bootstrapped.

Slightly different from the above work, in studying tests of linear regression model disturbances, Dubbelman et al. (1978) considered the typical spectral shape of economic time series. This would be important in cases where artificially generated regressions were used in Monte Carlo studies of the properties of such tests. It seems that the vast majority of economic variables investigated, after removal of trend and seasonal components, have similarly shaped power spectra, the typical shape being represented in Dubbelman et al. (1978) figure 2.

That is, the contribution of low frequency components to the variance of an economic time series is relatively high. One of the most obvious properties of a sample from a stochastic process having a spectrum of the typical shape is a visible long-term fluctuation which is not periodic. Comparisons of the above figure with that a first-order autoregressive scheme $x_i = \theta x_{i-1} + \eta_i \eta = (\eta_1,...,\eta_n) \sim$ N(0,I) suggests that the latter might generate typical economic time-series. Empirical investigations seem to approximate this conclusion, provided the autocorrelation parameter is given a very high value, e.g. 0.95 or higher, so that much weight is given to the low frequency components. This agrees with the idea that economic time-series are often almost unstable.

If, for example, $\theta = 0.95$, the height of figure 2 at frequency zero is 200/ π , whereas $\theta = 0.7$ for (the value used by L'Esperance and Taylor (1975)), it is $2/\pi$. Thus the power spectrum corresponding to the series generated by L'Esperance and Taylor differs very much from the power spectrum of economic time series even if trends are removed from the latter. The conclusions of L'Esperance and Taylor, accordingly, are misleading. Dubbelman et al. (1978) suggested either increasing the value of λ or adding a trend (or both) as a means of improving the design of L'Esperance and Taylor's simulations.

We are mainly concerned with a similar problem that may arise in Godfrey and Pesaran's (1983) study. The aim of our study is to investigate precisely this aspect of the MC experiment of GP[1]. Given that the regressors of the null model in GP were generated as serially and contemporaneously independent standard normal variates that is equivalent to the extreme case where the autocorrelation coefficient was zero for all those regressors. In terms of Dubbelman et al. (1978) designation, these cannot be regarded as being representative of typical economic phenomena.

A small MC experiment with modified parameters in the generating mechanism for the regressors in the null model is carried out. To enable comparison with the results of GP, we consider replicating their experiment in Table 1. This may reveal whether the nature of the regressor they have used had affected their results. We shall generate our regressor variables as typical economic time-series, according to the recommendations of Dubbelman et al. (1978).

[1] Hereafter, Godfrey and Pesaran (1983) will be abbrebiated GP

6.2. TEST STATISTICS

The test statistics considered in GP were and GP subscripted their test statistics with 0 and 1 to indicate that the null (H_0) and alternative hypotheses (H_1) were being tested respectively. These are given by

$$W_0 = \frac{(n-k_1)(\tilde{\sigma}_1^2 - \tilde{\sigma}_{10}^2)}{\{2\tilde{\sigma}_0^4 tr(B^2) + 4\tilde{\sigma}_0^2 e_{100}' e_{100}\}} \tag{1}$$

$$\tilde{N}_0 = \frac{\frac{1}{2}(n-k_1)\log\left[\frac{\tilde{\sigma}_1^2}{\tilde{\sigma}_{10}^2}\right]}{\left\{\left[\frac{\tilde{\sigma}_0^2}{\tilde{\sigma}_{10}^4}\right]\left[e_{100}' e_{100} + \frac{1}{2}\tilde{\sigma}_0^2 tr(B^2)\right]\right\}^{\frac{1}{2}}} \tag{2}$$

$$J_0^2 = \frac{(n-k_0-1)(e_0'\hat{y}_1)^2}{(e_0'e_0)(e_{01}'e_{01}) - (e_0'y_1)^2} \tag{3}$$

$$JA_0^2 = \frac{(n-k_0-1)(e_0'e_{10})^2}{(e_0'e_0)(e_{100}'e_{100}) - (e_0'e_{10})^2} \tag{4}$$

$$F_0 = \frac{(y'\overline{P}_X y - y'\overline{P}_{x,\,z} y)/k_1}{y'\overline{P}_{x,\,z} y/(n-k_0-k_1)} \tag{5}$$

The test statistics corresponding to H_1 follow by interchanging the subscripts 0 for 1 in (1) to (5) above.

The first two are adjusted Cox-type test statistics derived in GP and are asymptotically equivalent. Equation (3) gives the Davidson-MacKinnon J-test for H_0 which may be calculated as a test of $\alpha = 0$ in the artificial regression $y = Xb_0 + \alpha[P_Z y] + u$ under H_0, this is asymptotically distributed as a standard normal deviate. Equation (4) gives the Fisher-McAleer JA – test for H_0, which can be calculated as a test of $\zeta = 0$ in the artificial regression $y = Xb_0 + \zeta[P_z Xb_0] + u$. Equation (5) gives the classical F – statistic for the test in the combined regression $y = Xa + Zb + u$, in accordance with the comprehensive model-testing approach.

The adjusted Cox-type statistics based on mean- and variance-adjustments to standard Cox-type tests were proposed with the intention of overcoming the well-known problem of standard Cox-type tests having small sample significance levels, which consistently exceed the nominal level. The adjustments proceed from the observation that a possible reason for the unsatisfactory small-sample performance of the conventional (unadjusted) Cox-type test statistic (denoted by T_0) is the fact that the expected value of $n^{-1/2}T_0$ is zero under the null hypothesis only asymptotically. In small-samples, this will not be true. As GP noted, simple adjustments to ensure that the mean is zero are difficult to obtain in view of the logarithmic form of T_0, as seen in

$$T_0 = \frac{n}{2}\log\left\{\frac{e_1'e_1}{e_0'e_0 + e_{10}'e_{10}}\right\}.) \qquad (6)$$

It is obviously much easier to work with linearised forms and that is precisely what GP does in deriving W_0. McAleer (1987) showed that the test JA_0^2 is an exact test, being distributed as the $F(1, n - k_0 - 1)$ distribution under H_0.

On the other hand, the JA_1^2 is not exact under H_0. He also showed that the only separate test, which is exact for both models in a paired comparison, is the standard F test based on a comprehensive model such as that which we have in (5). These observations will be important in evaluating the results of any simulation.

Review of Godfrey and Pesaran's Result

GP considered the test statistics W, $\tilde{N}$, JA, J and F - distinguishing between the test for H_0 and through the use of subscripts, as mentioned above.

[2] McAleer (1987, pg.165) showed that a test which in not exact for the null can, under certain circumstances be made exact for the alternative through straightforward variance adjustments. As his lucid analysis showed, the JA test is quite amenable to such variance.

There were four sets of experiments in their study to evaluate the small-sample performance of the above tests, intended to cover "various possibilities of interest to applied econometricians". Experiments were distinguished on the basis of:

- The coefficient of determination, R^2, which determined the value which the error variance was to take;
- The simple correlation between the x_{ti} and $z_{ti,}$ set equal to ρ for all i where $i = 1, \ldots, \min(k_0, k_1)$;
- The simple size, n, set to equal 20,40 or 60 as a means of evaluating the quality of asymptotic approximations;
- The difference between the number of regressors in the null model $k_{0,}$ and that under the alternative, k_1;
- The error distribution, as a means of determining robustness to non-normality;
- Presence or absence of a lagged dependent variable.

In the first set of experiments, consequences of having an unequal number of regressors under the null and alternative hypotheses were considered. The second set of experiments covered situations where there were equal numbers of regressors. The third concentrated on variations in error distributions, and the fourth considered the case where dynamic specification played a part. Throughout their experiments, the nominal critical value was set at 5 per cent.

Their results for 500 replications showed that the *J* test was a dismal performer- in general- and the worst among the tests, having finite sample significance levels, which exceeded the nominal to an extent not seen in the others. Hence, it was left out in the results presented. GP also found the performance of the *JA* test to be unsatisfactory compared to the *W*, $\tilde{N}$ and *F* tests in terms of power when the true model has more regressors than the false model. Hence, they presented results mainly for these latter three tests. They found the adjusted Cox-type tests to be effective in removing the tendency of over-rejection of the null hypothesis of traditional Cox-type tests. In fact, the W^2-test had a slight tendency to under-reject. They found, as expected, that powers of the four tests fell with n and R^2. For moderate values of these parameters, they concluded, "the $F-$test is usually inferior in terms of power with the superiority of $\tilde{N}$ and W not being bought at the price of over-rejection of the true model" (Godfrey and Pesaran, pg.144), hereafter GP.

6.3. Experimental Outline

It seems unnecessary to consider distributional forms other than the normal, in view of assertion that it is the extent of non-normality in the regressors that determines sensitivity to non-normality in the dependent variable. This implies that merely varying the distributional form of the latter would not constitute a significant experimental variation, and hence would not be useful. Our limited study therefore involves re-doing the experiment corresponding to Table 1 in GP with the revised regressor-generation mechanism described above. (This means that we have ignored the J test totally.)

Generation of Artificial Regressors

Under the null hypothesis, the model used to generate the n observations on the dependent was

$$y_t = \sum_{t=1}^{k_0} x_{ti} + u_t, t = 1,2,\cdots,n,$$

Where for each replication, the values of x_{ti} were generated by an *AR(1)* process

$$x_{ti} = \vartheta_i x_{t-1,i} + v_{ti}$$

With v_{ti} serially and contemporaneously uncorrelated, and generated as standard normal variates using the subroutine DISTI from inder's DYNAM.FOR. Values of $\vartheta_i \{i = 1,...,k_0)$ have to be 'high' i.e. close to one. For our case, we set $\vartheta_i = \vartheta = 0.95$ for all i.

Under the alternative H_1, the explanatory variables were generated as in GP (pg.141) i.e. via

$$z_{ti} = \lambda_i x_{ti} + \varepsilon_{ti}, \qquad i = 1,2,\cdots,k_1,$$

With ε_{ti} also serially and contemporaneously uncorrelated among themselves and with the v_{ti}'s where the values of λ_i were set according to

$$\lambda_i = \begin{cases} \dfrac{\rho}{\left(1-\rho^2\right)^{1/2}}, & i = 1,2,\cdots,\min(k_0, k_1) \\ 0 \quad , & i = k_0 + 1,\cdots,k_1 \ if \quad k_0 > k_1, \end{cases}$$

To ensure that the simple correlation between $x_{ti}h$ and z_{ti} is set equal to ρ for i=1,2,..., min(k_0,k_1) and other x_{ti} and z_{ti} are uncorrelated.

As a control and a check on our computations, we recalculated GP's Table 1 with $\vartheta = 0$. We should obtain results closely similar to theirs.

The following section presents result of size and power calculations for 2000 replications in each experiment. In each replication, disturbances $u_t(t = 1, 2,\ldots,n)$ were generated using the subroutine DISTI from Inder's DYNAM.FOR, and the four test statistics given by (1), (2), (4), and (5) were calculated. Estimates of the significance levels[3] of these tests were computed as the proportion of times that their corresponding test statistic exceeded the appropriate 5% critical value.

The critical values for the tests were calculated using the IMSL subroutines MDCHI and MDFI. These calculate the significance values for given levels of significance for the central - χ^2 and central –F distributions respectively. In the case of the test W_0^2 and N_0^2 statistics-, which are, squared standard normal deviates – the critical values were obtained from the χ_1^2 distribution. In the case of the JA_0^2 test statistic, the distribution was used; and in the classical F_0 test statistic, the $F(k_1, n-k_0-k_1)$distribution was used.

Classical econometric principles applied in the case of nested hypothesis-testing imply that consideration of the proportion of times the null is rejected when the alternative is the true data generating process leads to an acceptable indicator of power. Godfrey and Pesaran (1983, pg.143, GP here after)

[3] This is coded in GP1.FOR; input file is GP1IN.DAT and output file is GP1.OUT.

Table I. Estimates of power and size from 2000 replications

$$k_0 = 4, \ R^2 = \rho^2$$

		ESTIMATED		POWER		ESTIMATED SIZE			
R^2	N	F	JA^2	$\tilde{N}^2$	W^2	F	JA^2	$\tilde{N}^2$	W^2
.3	20	24.50	26.90	38.20	32.30	5.10	4.65	4.05	2.65
	40	57.35	59.59	70.95	67.60	5.30	4.50	4.40	3.35
	60	80.20	81.75	88.45	88.05	5.35	5.50	4.80	4.40
.5	20	51.80	61.70	75.50	68.00	5.10	5.25	4.45	3.30
	40	9.00	92.70	93.70	94.30	5.30	5.05	4.45	3.75
	60	95.70	95.20	95.05	95.90	5.35	5.10	4.75	4.95
.8	20	94.30	94.75	93.50	94.95	5.10	5.30	4.75	4.45
	40	94.95	95.60	94.50	95.55	5.30	5.50	4.80	4.80
	60	95.59	94.95	94.50	95.35	5.35	5.20	5.10	4.95
.85	20	94.85	95.10	93.70	94.95	5.10	5.45	4.65	4.55
	40	94.95	95.50	94.90	95.55	5.30	5.65	4.70	5.00
	60	95.50	95.30	94.55	95.15	5.35	5.20	5.00	5.05
.9	20	94.80	95.20	93.5	95.05	5.10	5.40	4.55	4.75
	40	94.90	95.05	94.95	95.50	5.30	5.70	5.20	5.20
	60	95.50	95.00	94.75	95.20	5.35	5.30	5.10	5.05

Notes: (a) All estimates are presented as percentages to facilitate comparison with GP's (pg.145) Table 1.

b) Values in bold indicate best power for given $R^2$and n.

(c) Values in italics indicate worst power for given $R^2$and n.

translated these same principles to the tests of non-nested hypotheses and used the probability of making a correct decision (i.e. accepting the true model and rejecting the false one) as the appropriate indicator of power. We therefore estimated the power of our tests as the proportion of times where both the null is rejected (based on one of F_0, W_0^2, $\tilde{N}_0^2$ or JA_0^2 exceeding the relevant critical value for tests of the alternative)[4]. All computations were carried out in Fortran using double precision arithmetic[5].

[4] These computations are coded in GP2.FOR. it is obvious that critical values for tests of the alternative hypothesis are the 95 percentile values in the relevant distributions as follows: the χ_1^2 in the case of the W_1^2 and $\tilde{N}_1^2$; the $F(1, n-k_1-1)$ In the case of the JA_1^2; and the $F(k_0, n-k_0-k_1)$ in the case of F_1.

[5] This was the source of a major problem in the computations for the MC experiments. Initially, standard normal variates were generated using the IMSL subroutine GGNML. However, theses weren't suitable for double-precision work and a major debugging exercise revealed

6.4. Results

6.4.1. Size of the Tests

Table I gives the estimated sizes for the experiments tabulated in GP's Table 1. As noted in section 2, the *F* test in the comprehensive model is an exact test. The above results are therefore consistent with this, all observable discrepancies falling within the bounds of acceptable sampling variation. As can be anticipated from the structure in (5), the *F* statistic is invariant to the disturbance, which is effectively what is begin varied in incrementing the coefficient of determination in Table I. As in GP's Table 1, we find the W^2 -test has a slight tendency to over-adjust for the previously noted finite-simple over-rejection rates of traditional Cox-type tests. We also found the same tendency to under-reject in the case of $\widetilde{N}^2$. This minor deviation from GP should be attributable to our more rigorous practice of using 2000 replications[6]. With our modified autoregressive parameter in the generation of the null model regressors (0.95 instead of 0), we find that estimates of size seem to have maintained roughly the same order of agreement with the nominal size I the case of the modified Cox-type tests. In terms of maximum deviation from the 5% figure, there is a barely discernible improvement in this case for these two tests. Though there is still the same tendency to under-reject. We also find the close adherence of estimated sizes to nominal ones in the case of the *F* statistic, as expected. The predominant concern with regard to the estimated size has been to establish a common ground for comparison among the tests. In both Table I and II, this has assuredly been achieved given the generally close adherence of estimated sizes to the nominal 5% level. We can then go on to make justifiable power comparisons.

---

that the use of double-precision in the main program GP1.FOR was incompatible with the use of GGNML. Examination of the error vectors generated by GGNML in *our* experiments showed that where a vector of length 60 was concerned, the values for positions (approx.) twenty to thirty were very large – of the order of 10^8. Just as startling was the fact that those for positions thirty-one to sixty were zerol we therefore turned back to the subroutine DISTI in STATIC.FOR -a modified version of Inder's DYNAM.FOR – to general our standard normal deviate.

[6] The sampling error a nominal significance level of 5% when there are 2000 replications is 0.487. Inspection of Table I shows none of the estimated sizes of the two modified Cox-type tests exceeds the nominal size by more than 1 standard error while there are clearly a few falling below the nominal size by more than 1 standard error.

6.4.2. Power of Tests

As mentioned above, the power defines as the "probability of making the correct decision", i.e. of accepting the true model and rejecting the false one. It is evident from Table I that even in our recalculations of the same experiments in GP with all parameters unaltered (i.e. $\vartheta = 0.0$), the estimated powers of the JA^2 test are not as poor as they suggest. In every single case in Table I, estimates of power rose dramatically in comparison to the other three tests. The resulting profile establishes the JA^2 Test as a satisfactory performer in terms of finite-sample power and size relative to the W^2, $\widetilde{N}^2$ and F test.

In Table II, the estimates of power were in general higher than those in Table I. This is easily seen by comparing the worse and best power values for each set of (R^2, n) values. More interesting again is the acceptable and sometimes impressive performance of the JA^2 test.

Table II. Estimates of power and size from 2000 replications

$$k_0 = 4, R^2 = \rho^2$$

		ESTIMATED		POWER		ESTIMATED SIZE			
R^2	N	F	JA^2	$\widetilde{N}^2$	W^2	F	JA^2	$\widetilde{N}^2$	W^2
.3	20	32.40	42.90	56.30	49.05	4.75	4.35	4.80	3.60
	40	58.85	69.75	78.10	71.75	4.95	5.35	4.75	4.15
	60	80.30	87.35	90.05	87.95	5.65	4.75	4.90	4.60
.5	20	66.60	78.95	87.80	81.60	4.75	4.25	4.95	3.75
	40	90.05	93.00	94.05	93.45	4.95	5.35	5.25	5.00
	60	94.75	95.20	94.80	95.30	5.65	4.90	4.65	4.55
.8	20	94.95	95.30	94.55	96.05	4.75	4.25	4.55	3.85
	40	95.20	94.35	94.35	94.80	4.95	5.25	5.10	4.85
	60	94.90	95.45	94.75	95.35	5.65	4.65	4.55	4.55
.85	20	95.05	95.25	94.85	95.85	4.75	4.20	4.15	3.90
	40	95.25	94.35	94.30	94.95	4.95	5.30	4.95	5.00
	60	94.95	95.45	94.85	95.40	5.65	4.60	4.45	4.55
.9	20	95.05	95.25	95.15	95.70	4.75	4.25	4.15	4.00
	40	95.10	94.45	94.15	95.00	4.95	5.25	5.05	5.10
	60	95.05	95.50	95.05	95.45	5.65	4.60	4.50	4.55

6.5. CONCLUSION

One of the aims of GP was to present modified Cox-type statistics intended to alleviate the problem of over-rejection of the true model. Their results showed that the W- and $\tilde{N}$ −tests have estimated significance levels which were reasonably close to the nominal levels. Our modified experiments reveal this aspect of their results had not been compromised by their use of non-typical regressor design.

The primary aim of this study has been to highlight the consequences which improper Monte Carlo design may have on comparisons among competing testing procedures, in the spirit of Dubbelman et al.'s (1978) paper. The results of this paper have reinstated the JA^2 test as a serious contender within the array of inferential tools in non-nested hypothesis testing. It appears however that the original suspicion that the regressor-generation mechanism may have affected the results significantly hasn't been confirmed. However, rather startling discrepancies between our relative power profile and that obtained in GP with exactly the same parameter values have been obtained. It is unclear at this stage as to the source of this problem. This will have to be confirmed with a more extensive study taking into account a larger range of design-parameter values.

Chapter 7

SUMMARY AND CONCLUSION

7.1. INTRODUCTION

This book significantly contributes to the practical use of non-nested hypothesis testing when the null (H_0), alternative (H_1), or both hypotheses are specified as errors-in-variables models (EVM). The study is primarily motivated by a recent surge in applied econometrics to hypothesize any given economic phenomenon into two or more separate models. Two models are formally termed separate or non-nested if neither one can be obtained as a limiting or special case of the other model. Two linear regression models are said to be globally non-nested if they contain no common regressors; on the other hand, two competing models composed of both common and distinct variables are partially non-nested.

7.2. SUMMARY

The behavior of standard non-nested tests and IV-modified non-nested tests is now summarized.

7.2.1. Standard Non-Nested Tests

In general, the likelihood ratio test follows an intractable sampling distribution when the null hypothesis is non-nested; the asymptotic powers of the conventional F-, Wald, and Lagrange Multiplier tests decline with increasing number of distinct regressors between a non-nested H_0 and H_1.

Thus, following the pioneering work by Cox (1961, 1962), several specialized tests involving a modified likelihood ratio approach by Cox or the artificial embedding approaches by Atkinson (1970) and Davidson and MacKinnon (1981a) have been developed to test non-nested single or multivariate regression models. Both approaches provide consistent and asymptotically equivalent statistics for testing the null hypothesis against single or multiple alternatives. Using the encompassing principle, where a 'highly parameterized model is constructed to encompass the parsimoniously parameterized non-nested alternatives, Mizon (1984) and Mizon and Richard (1986) obtain the slope, variance, and complete parametric encompassing tests which are also consistent and asymptotically equivalent to the Cox test and linearly embedded tests. For linear regression models, the complete parametric encompassing test can be obtained as the Wald or conventional F-test from a compound model that nests or encompasses the non-nested H_0 and H_1.

The asymptotic powers of the Cox type tests and linearly embedded tests are the same under local and fixed alternatives. However, the local power of these tests exceeds that of the conventional F-test when H_0 and H_1 have more than one non-overlapping variable; under the fixed alternative approach, the powers of these tests cannot be unambiguously ranked. Using Bahadur's (1960) criterion, Pesaran (1984) concludes higher efficiency for a specific form of the Cox test relative to the F- and Davidson and MacKinnon's (1981a) J-test. However, this result cannot be generalized to other forms of the Cox test, and his approach is inappropriate when the competing hypotheses are partially non-nested. Thus, the analytical evidence fails to isolate a uniformly-most-powerful test among the existing non-nested tests. Nonetheless, linearly embedded tests are sometimes preferred in empirical applications for their computational simplicity.

Most of the specialized non-nested tests have only asymptotically correct sizes; the F-test and Fisher and McAleer's (1981) JA-test have exact sizes. Monte Carlo studies suggest that the Cox and J-tests experience an over-rejection of the true null hypothesis in small samples (in some instances, this over-rejection can be quite severe). Factors contributing to this over-rejection are a poor fit of null hypothesis, low colinearity between the regressors of H_0 and H_1, and a false model having more regressors than the true model. The power of the JA-test is generally lower than the Cox test when the number of regressors in the competing models is different.

Typically, these non-nested tests deal with separate regression models under the full ideal conditions and are, therefore, based on the least squares or maximum likelihood estimators of the competing models. However, empirical studies generally involve one or more violations of the classical assumptions which may be serious enough to plague the otherwise correct behavior of non-nested tests. For example, Pesaran and Deaton (1978) note the lack of robustness of the Cox test to autocorrelation; McAleer et al. (1982) show the inconsistency of non-nested tests when some of the regressors are incorrectly omitted.

The effect of applying these standard non-nested tests to mismeasured regression models has been studied in this book and the following results were formally demonstrated.

(i) Under the null EVM, the Cox type tests, linearly embedded tests, comprehensive model tests such as the F-test, Dastoor's (1983) R-test, and various forms of the encompassing tests converge asymptotically to non-central limits. These non-central values involve unidentified parameters so that no mean adjustments are typically possible. Thus, the null EVM based on standard non-nested test will be rejected asymptotically regardless of whether it is "true" or not.

(ii) However, the existing non-nested tests remain consistent so that a false null, whether an EVM or correctly measured model, is rejected with certainty (asymptotically) by either an EVM or a correctly measured alternative.

(iii) The Cox, J-, JA-, and Wald encompassing tests cannot be proved to be asymptotically equivalent nor ranked in terms of their asymptotic means when the null hypothesis is an EVM.

(iv) These distortions apply only to regression models with mismeasured regressors. The standard non-nested tests are asymptotically correct and well-behaved when only the regress and is mismeasured.

Recognizing this failure of conventional non-nested tests to ascertain the validity- of an EVM, several modified tests are proposed in this book. The description and analysis of these modified tests are now summarized.

7.2.2. Modified Non-Nested Tests

The inconsistency of least squares estimators for an EVM is presumably the basic cause of the asymptotic divergence between the nominal and true sizes of the conventional non-nested tests of a null EVM. For this reason, the least squares estimators are replaced with the consistent IV estimators of the mismeasured model(s) and various tests are modified accordingly. The following IV-based, non-nested tests have been proposed in the literature:

a) Godfrey's (1983) G- and T-tests as IV analogues for the Cox and JA-tests,
b) Ericsson's (1983) numerical equivalent of the G-test and an IV-based Wald encompassing test,
c) The J-type test of MacKinnon et al. (1983) involving two stage least squares estimation of an artificially-embedded regression model,
d) Dastoor and McAleer's (1985) R-test (or asymptotically equivalent F-test) based on the LM and Wald principles.

The existing analysis of these tests relies on the use of a common yet unidentified, IV matrix for H_0 and H_1, which is presumably composed of regressors outside of H_0 and H_1. In addition, several of these existing tests are inappropriate when applied to IV estimation and testing of mismeasured regression models. To enhance the practical ability of the modified tests, a novel approach for the selection of IV matrices in a non-nested framework is proposed in this study. The following analytical results were shown to apply to the IV-modified tests of mismeasured regression models.

(i) Suppose initially that only one model, say H_0, is mismeasured and that H_1 is correctly measured. In this instance, replacing the least squares estimators in H_0 with their IV estimators in the Cox or variance encompassing tests does not provide an asymptotically zero mean test of the null EVM, and these limiting values consist of the unidentified variance-covariance matrix of measurement errors. Thus, for the non-nested EVM, these simple IV modifications fail to yield asymptotically correct sizes for these tests. Similarly, the G-test and IV-analogues of the Cox test also fail to ascertain the asymptotic truth of a null EVM if the alternative model is correctly measured.

(ii) Substituting the IV estimators for least squares estimators of the mismeasured model(s) in the J- and JA-tests, however, provides asymptotically zero mean tests ($\widetilde{J}$ and $J\widetilde{A}$, named IV-based tests) for a mismeasured H_0. Both of these tests are consistent and their fixed powers asymptotically equal that of the G-test. However, these tests cannot be computed or interpreted as the t-ratios from artificial regression models and the computational ease of the embedding approach is lost.

(iii) The statistics $\widetilde{J}^{m}$ and $J\widetilde{A}^{m}$ (termed IV-modified tests) are computed as the t-ratios from IV estimation of the augmented regression models. These tests are consistent and also asymptotically normal zero-mean tests of a mismeasured H_0. Unlike MacKinnon et al. and Godfrey, different IV matrices may be used for each model, and the augmented regressors, $\widetilde{y}_1$ or $\widetilde{\widetilde{y}}_{10}$, act as their own instruments. (Here $\widetilde{y}_1$ represents the vector of IV-predicted values for H_1 and $\widetilde{\widetilde{y}}_{10}$ is the vector IV-predicted values when $\widetilde{y}_0$ (IV-predicted values for H_0) is regressed on the regressors of H_1.) Under this IV regime, however, it is not possible in general to rank either the local or fixed powers of these tests.

(iv) Similarly, when the modified slope parametric encompassing test consists of separate IV matrices for H_0 and H_1 (or compound model IV is constructed by using non-overlapping variables from each model), the resulting statistic is an asymptotically correct test of H_0 whose local power is less than that of the G-test when the number of distinct regressors in H_1 exceeds one. However, the fixed powers of the two tests cannot be ranked.

This study also explores in some detail the option of choosing IV matrices from within one or both of the non-nested models. It was found that the following limitations apply to such a choice.

a) If only one of the non-nested models, say H_0, is the EVM and the regressors in H_1 are chosen as the instruments for H_0, then $\widetilde{J}$, $J\widetilde{A}$, $\widetilde{J}^{m}$, $J\widetilde{A}^{m}$ and the encompassing tests fail to exist. In general, it implies that the IV matrix of H_0 and the regressors of H_1 should be linearly independent of each other.
b) However, when both non-nested models are mismeasured, the regressors of one model may be selected as the IV for the other model if the number of distinct regressors in each model exceeds or equals the number of mismeasured regressors in other model.
c) The existence of various modified tests requires that the IV matrix for each model does not include all regressors of the other model.
d) If a distinct, mismeasured regressor is chosen as the IV for the mismeasured regressor of the other model then the encompassing test

is an asymptotically non-central test and the non-centrality parameter will be unidentified.

e) In the special case that H_0 and H_1 use the same IV matrix which is composed of the fixed regressors of both models, the G- and encompassing tests do not exist and $\widetilde{J} = J\widetilde{A}$ and $\widetilde{J}^m = J\widetilde{A}^m$.

Finally, the book provides first hand Monte Carlo evidence of the behavior of modified non-nested tests when the IV for both models is selected according to choice (e) above.

The design of the true DGP and the fixed alternative hypothesis is in line with Godfrey and Pesaran (1983). However, the introduction of mismeasured regressors in both models, followed by instrumental variable estimation, is the distinct feature of this experiment.

Several measures of performance evaluation for the modified tests $\widetilde{J}$ and $\widetilde{J}^m$ are computed in 320 possible combinations of the sample size (n), the coefficient of multiple determination (R^2) for H_0 , the correlation coefficient (ρ^2) between the regressors of H_0 and H_1 , the number of distinct (k) and mismeasured regressors (m), and the measurement error variance (σ_m^2).

Several interesting results are obtained from this specialized experiment. For all combinations of R^2 and ρ^2 , both tests approach the correct size and have increasing power functions as the sample size increases. As expected, the powers of both tests vary directly with R^2 and inversely with k and σ_m^2. However, the effect of ρ^2 on the power of these tests is twofold. An increase in ρ^2 from 0.30 to 0.50 causes power to increase whereas a further increase in leads to decreased power, *ceteris paribus.*

The possible explanation for the former effect deals with the improved efficiency of the IV estimators with increased values; on the other hand, the latter effect involves increased closeness between the null and fixed alternative hypothesis so that the test statistics are unable to differentiate between H_0 and H_1. Finally, the powers of both tests are significantly higher when n = (60, 100), R^2 = (0.99, 0.95, 0.80) and ρ^2 = (0.30, 0.50); $\widetilde{J}$ is usually less powerful than $\widetilde{J}^m$.

7.3. Conclusion

This book resolves some practical issues in the modelling of economic and other social science phenomena when the sample data are mismeasured and separate theories explain a given economic or social phenomenon. In such a separate or non-nested framework, if a priori evidence suggests some measurement problems for the dependent variable whereas the explanatory variables in all non-nested models are non-stochastic, then the conventional non-nested tests can be readily used to verify the truth of these separate models. However, when the regressor(s) in one of the two non-nested models are diagnosed to be mismeasured then the use of conventional non-nested tests provides one of the following two outcomes asymptotically:

a) Reject both non-nested models as the true models,
b) Reject the mismeasured model(s) and accept the correctly measured model as a true model.

The first outcome is misleading since a true (mismeasured) model is rejected by an incorrectly applied conventional non-nested test. However, the second outcome reaches the right conclusion since the correctly measured null model (if true) cannot be rejected by the mismeasured alternative model. Thus, the practitioner confronted with the first outcome or dealing with two non-nested, mismeasured models should use the modified non-nested tests to guarantee correct inference. In the case of a mismeasured model tested against a correctly measured alternative, it is extremely important to ensure linear independence between the IV matrix and the correctly measured regressors; otherwise, the modified tests will collapse. When both non-nested models are mismeasured, one can typically use the simple and practical IV selection procedure proposed in this study. Finally, among the various modified tests, the computational convenience of $\widetilde{J}^{m}$, supplemented with its higher power, as observed by Monte Carlo experiments, enhances its practical utility for a variety of real world applications.

APPENDIX A

The following table summarizes the experimental design parameter values reported in the tables in appendix B. Its purpose is to permit ease of comparison between design points.

Graphs based on table 1 data

Non-nested Regression

R^2 Vs Estimated Power

Table 1-a **Table 1-a** N=20

R^2	F	JA^2	$N\sim^2$	W^2
0.30	24.50	26.90	38.20	32.30
0.50	51.80	61.70	75.50	68.00
0.80	94.30	94.75	93.50	94.95
0.85	94.85	95.20	93.50	95.05
0.90	94.60	95.20	93.50	95.05

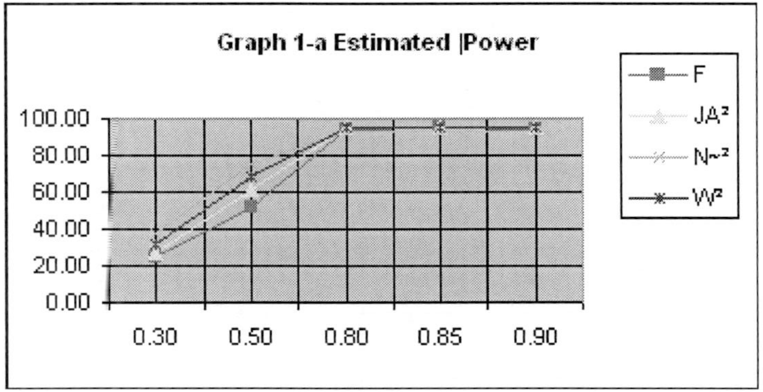

Table (Continued)

Table 1-b

Table 1-b	n=40	Power		
R^2	F	JA^2	$N\sim^2$	W^2
0.30	57.35	59.59	70.95	67.60
0.50	90.00	92.70	93.70	94.30
0.80	94.95	95.60	94.50	95.55
0.85	94.95	95.50	94.90	95.55
0.90	94.90	95.05	94.95	95.50

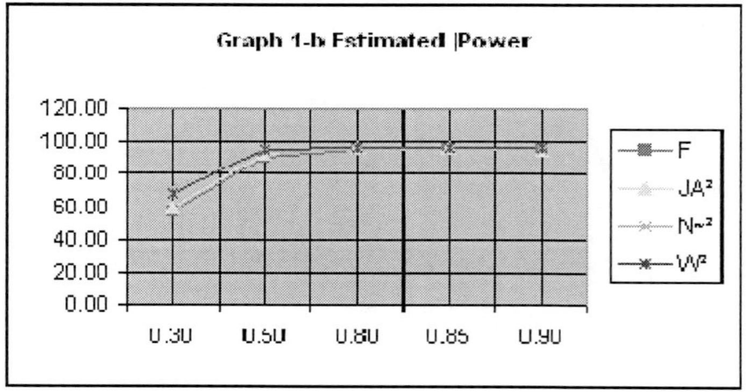

Table 1-c

Table 1-c	n=60			
		Power		
R^2	F	JA^2	$N\sim^2$	W^2
0.30	80.20	81.75	88.45	88.05
0.50	95.70	95.20	95.05	95.90
0.80	95.59	94.95	94.50	95.35
0.85	95.50	95.30	94.55	95.15
0.90	95.50	95.00	94.75	95.20

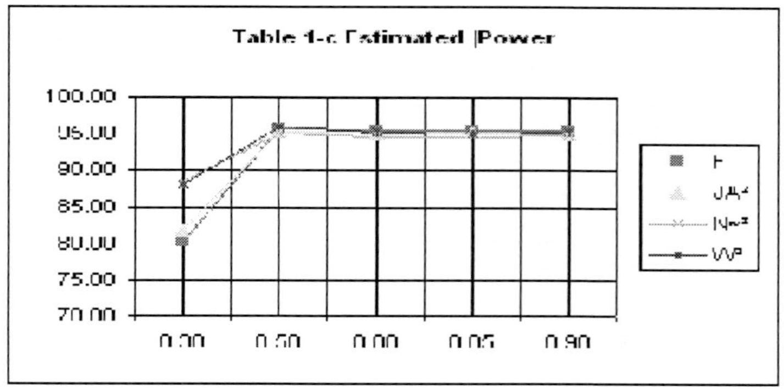

Table 1-d

Table 1-d	n=20	estimated size		
R^2	F	JA^2	$N\sim^2$	W^2
0.30	5.10	4.65	4.05	2.65
0.50	5.10	5.25	4.45	3.30
0.80	5.10	5.30	4.75	4.45
0.85	5.10	5.45	4.65	4.55
0.90	5.10	5.40	4.55	4.75

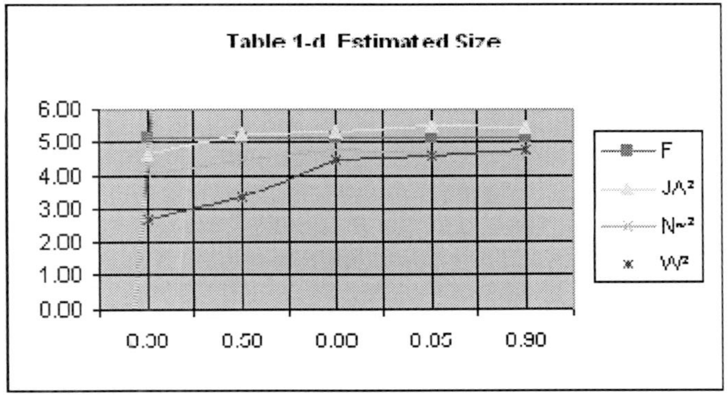

Table 1-e	n=40	estimated size		
R^2	F	JA^2	$N\sim^2$	W^2
0.30	5.30	4.50	4.40	3.35
0.50	5.30	5.05	4.45	3.75
0.80	5.30	5.50	4.80	4.80
0.85	5.30	5.65	4.70	5.00
0.90	5.30	5.70	5.20	5.20

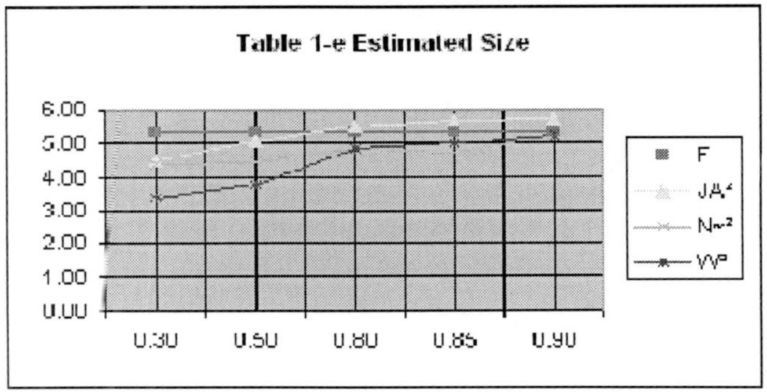

Table (Continued)

Table 1-f	n=60	estimated size		
R^2	F	JA^2	$N\sim^2$	W^2
0.30	5.35	5.50	4.80	4.40
0.50	5.35	5.10	4.75	4.95
0.80	5.35	5.20	5.10	4.95
0.85	5.35	5.20	5.00	5.05
0.90	5.35	5.30	5.10	5.05

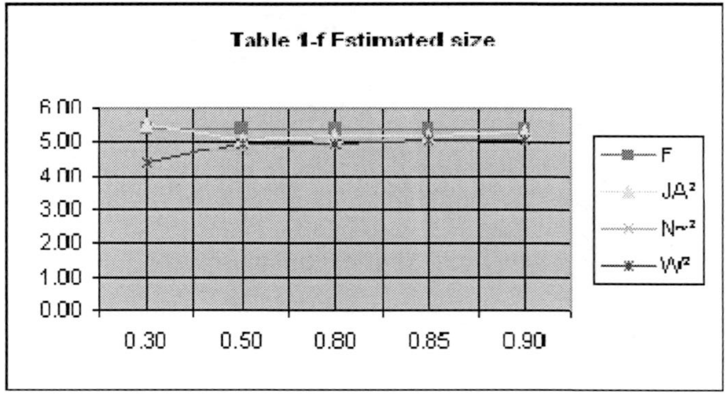

Graphs based on table 2 data

Table 2-a	Power	n=20		
R^2	F	JA^2	$N\sim^2$	W^2
0.30	32.40	42.90	56.30	49.05
0.50	66.60	78.95	87.80	81.60
0.80	94.95	95.30	94.55	96.05
0.85	95.05	95.25	94.85	95.85
0.90	95.05	95.25	95.15	95.70

Graph 2-a Estimated Power

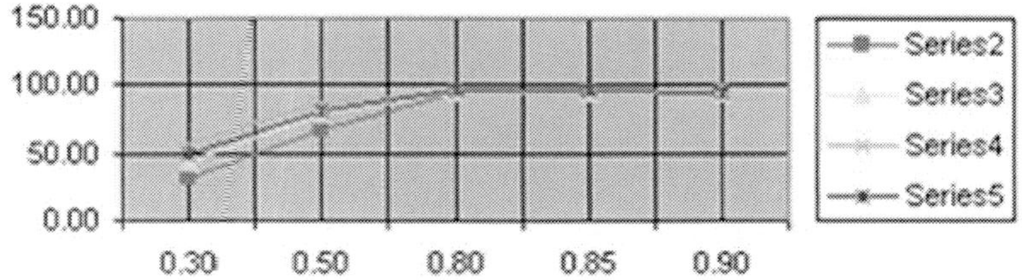

Table 2-b	Power	n=40		
R^2	F	JA^2	$N\sim^2$	W^2
0.30	58.85	69.75	78.10	71.75
0.50	90.05	93.00	94.05	93.45
0.80	95.20	94.35	94.35	94.80
0.85	95.25	94.35	94.30	94.95
0.90	95.10	94.45	94.15	95.00

Table 2-b Estimated Power

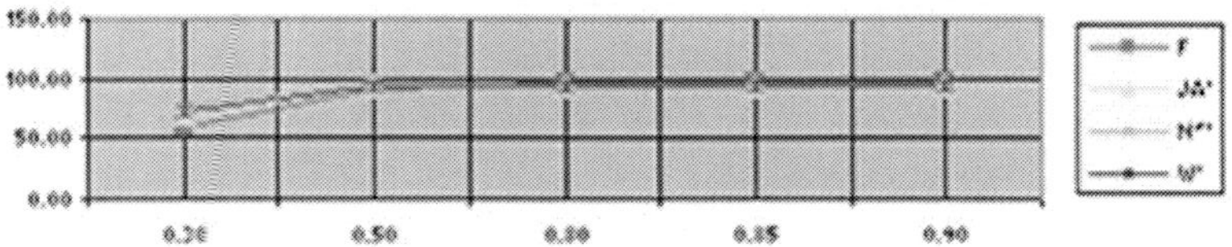

Table (Continued)

Table 2-c	Power	n=60		
R^2	F	JA^2	$N\sim^2$	W^2
0.30	80.30	87.35	90.05	87.95
0.50	90.05	93.00	94.05	93.45
0.80	95.20	94.35	94.35	94.80
0.85	95.25	94.35	94.30	94.95
0.90	95.10	94.45	94.15	95.00

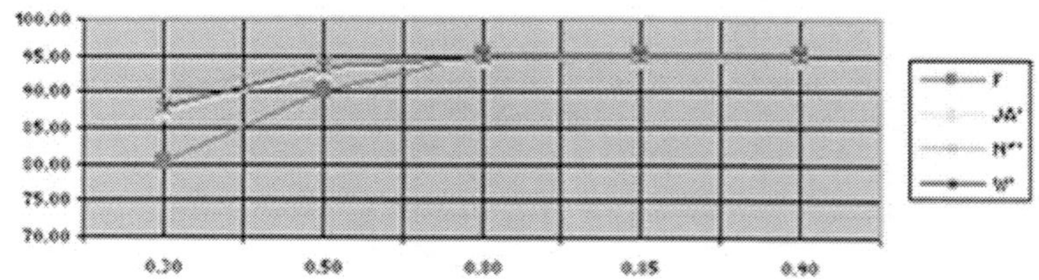

Table 2-d	Size	n=20		
R^2	F	JA^2	$N\sim^2$	W^2
0.30	4.75	4.35	4.80	3.60
0.50	4.75	4.25	4.95	3.75
0.80	4.75	4.25	4.55	3.85
0.85	4.75	4.20	4.15	3.90
0.90	4.75	4.25	4.15	4.00

Table 2-d Estimated Size

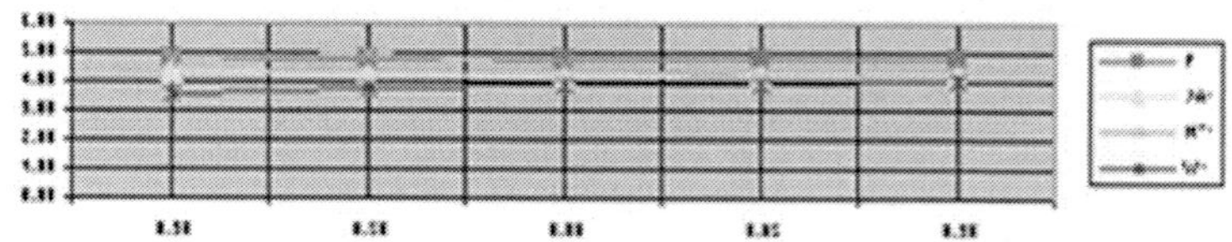

Table 2-e	n=40	estimated size		
R^2	F	JA^2	$N\sim^2$	W^2
0.30	4.95	5.35	4.75	4.15
0.50	4.95	5.35	5.25	5.00
0.80	4.95	5.25	5.10	4.85
0.85	4.95	5.30	4.95	5.00
0.90	4.95	5.25	5.05	5.10

Table 2-e Estimated Size

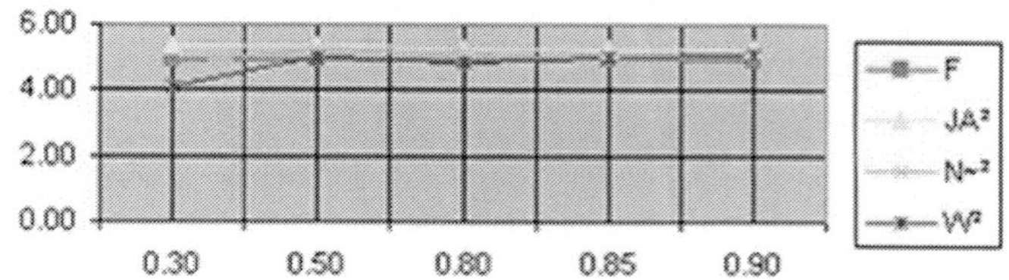

Table 2-f	n=60	estimated size		
R^2	F	JA^2	$N\sim^2$	W^2
0.30	5.65	4.75	4.90	4.60
0.50	5.65	4.90	4.65	4.55
0.80	5.65	4.65	4.55	4.55
0.85	5.65	4.60	4.45	4.55
0.90	5.65	4.60	4.50	4.55

Table 2-f Estimated Size

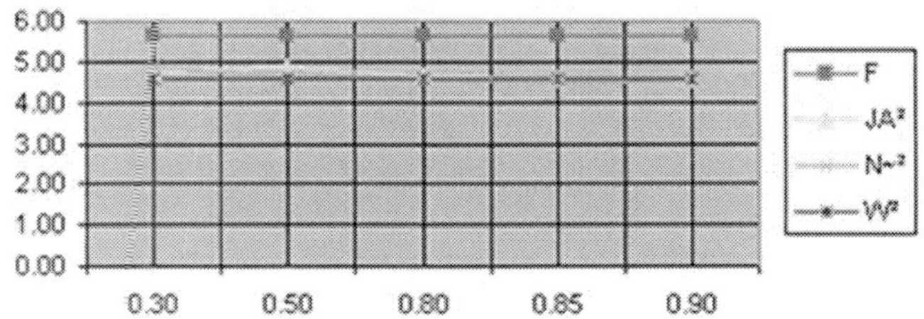

APPENDIX B

The Results of Monte Carlo experiment are presented in the following tables.

Table	k	m	σ_m^2	R^2	ρ^2
5.1.1	2	1	1.00	0.99	0.30
5.1.2	2	1	1.00	0.95	0.30
5.1.3	2	1	1.00	0.80	0.30
5.1.4	2	1	1.00	0.50	0.30
5.1.5	2	1	1.00	0.30	0.30
5.1.6	2	1	1.00	0.99	0.50
5.1.7	2	1	1.00	0.95	0.50
5.1.8	2	1	1.00	0.80	0.50
5.1.9	2	1	1.00	0.50	0.50
5.1.10	2	1	1.00	0.30	0.50
5.1.11	2	1	1.00	0.99	0.80
5.1.12	2	1	1.00	0.95	0.80
5.1.13	2	1	1.00	0.80	0.80
5.1.14	2	1	1.00	0.50	0.80
5.1.15	2	1	1.00	0.30	0.80
5.1.16	2	1	1.00	0.99	0.95
5.1.17	2	1	1.00	0.95	0.95
5.1.18	2	1	1.00	0.80	0.95
5.1.19	2	1	1.00	0.50	0.95
5.1.20	2	1	1.00	0.30	0.95
5.2.1	4	2	1.00	0.99	0.30
5.2.2	4	2	1.00	0.95	0.30
5.2.3	4	2	1.00	0.80	0.30

Table (Continued)

Table	k	m	σ_m^2	R^2	ρ^2
5.2.4	4	2	1.00	0.50	0.30
5.2.5	4	2	1.00	0.30	0.30
5.2.6	4	2	1.00.	0.99	0.50
5.2.7	4	2	1.00	0.95	0.50
5.2.8	4	2	1.00	0.80	0.50
5.2.9	4	2	1.00	0.50	0.50
5.2.10	4	2	1.00	0.30	0.50
5.2.11	4	2	1.00	0.99	0.80
5.2.12	4	2	1.00	0.95	0.80
5.2.13	4	2	1.00	0.80	0.80
5.2.14	4	2	1.00	0.50	0.80
5.2.15	4	2	1.00	0.30	0.80
5.2.16	4	2	1.00	0.99	0.95
5.2.17	4	2	1.00	0.95	0.95
5.2.18	4	2	1.00	0.80	0.95
5.2.19	4	2	1.00	0.50	0.95
5.2.20	4	2	1.00	0.30	0.95
5.3.1	2	1	0.25	0.99	0.30
5.3.2	2	1	0.25	0.95	0.30
5.3.3	2	1	0.25	0.80	0.30
5.3.4	2	1	0.25	0.50	0.30
5.3.5	2	1	0.25	0.30	0.30
5.3.6	2	1	0.25	0.99	0.50
5.3.7	2	1	0.25	0.95	0.50
5.3.8	2	1	0.25	0.80	0.50
5.3.9	2	1	0.25	0.50	0.50
5.3.10	2	1	0.25	0.30	0.50
5.3.11	2	1	0.25	0.99	0.80
5.3.12	2	1	0.25	0.95	0.80
5.3.13	2	1	0.25	0.80	0.80
5.3.14	2	1	0.25	0.50	0.80
5.3.15	2	1	0.25	0.30	0.80
5.3.16	2	1	0.25	0.99	0.95
5.3.17	2	1	0.25	0.95	0.95
5.3.18	2	1	0.25	0.80	0.95
5.3.19	2	1	0.25	0.50	0.95
5.3.20	2	1	0.25	0.30	0.95
5.4.1	4	2	0.25	0.99	0.30
5.4.2	4	2	0.25	0.95	0.30
5.4.3	4	2	0.25	0.80	0.30

Table	k	m	σ_m^2	R^2	ρ^2
5.4.4	4	2	0.25	0.50	0.30
5-4.5	4	2	0.25	O 30	0.30
5.4.b	4	2	0.25	0.99	0.50
5.4.7	4	2	0.25	0.95	0.50
5.4.8	4	2	0.25	0.80	0.50
5.4.9	4	2	0.25	0.50	0.50
5.4.10	4	2	0.25	0.30	0.50
5.4.11	4	2	0.25	0.99	0.80
5.4.12	4	2	0.25	0.95	0.80
5.4.13	4	2	0.25	0.80	0.80
5.4.14	4	2	0.25	0.50	0.80
5.4.15	4	2	0.25	0.30	0.80
5.4.16	4	2	0.25	0.99	0.95
5.4.17	4	2	0.25	0.95	0.95
5.4.18	4	2	0.25	0.80	0.95
5.4.19	4	2	0.25	0.50	0.95
5.4.20	4	2	0.25	0.30	0.95

REFERENCES

Andrews, D.F. (1971), "A Note on the Selection of Data Transformations", *Biometrika*, 58, pg: 249-254.

Aneuryn-Evans, G. and Deaton, A. (1980), "Testing Linear versus Logarithmic Regression Models", *The Review of Economic Studies*, 47(1), pg. 275-291.

Arraiz, I., Drukker, D., Kelejian, H., and Prucha, I. (2010), "A spatial cliff-ord-type model with heteroscedastic innovations: small and large sample results", *Journal of Regional Science,* 50, pg. 592–614.

Atkinson, A.C. (1970), "A Method For Discriminating Between Models", *Journal of the Royal Statistical Society*, Series B (Methodological), 32(3), pg. 323-353.

Bahadur, R.R. (1960), "Stochastic Comparison of Tests", *The Annals of Mathematical Statistics*, 31(2), pg. 276-295.

Bahadur, R.R. (1967), "Rates of convergence of estimates and test statistics", *The Annals of Mathematical Statistics*, 38, pg. 303-324.

Barnard, G.A. (1959), "Control Charts and Stochastic Processes", *Journal of the Royal Statistical Society*, 21(2), pg.239-271.

Bean, C.R. (1981), "A New Approach to the Empirical Investigation of Investment Expenditures: A Comment", *Economic Journal,* Royal Economic Society, Vol. 91(361), pg. 104-05.

Berger J.O., Mortera J. (1999), "Default Bayes factors for non-nested hypothesis testing", *Journal of American Statistical Association,* 94, pg. 542-554.

Bernanke B.S., Reiss P. C., Bohn H. (1986), "Alternative non-nested specification tests of time series investment models", *Research paper No. 890*, Graduate School of Business, Stanford University, USA.

Berndt, E.R., and Savin, N.E. (1977), "Conflict Among Criteria for Testing Hypotheses in the Multivariate Linear Regression Model", *Econometrica*, 45, pg.1263-1278.

Bhatti, M.I. (2012), "Cluster Effects in Mining Complex Surveys", Nova, NY, USA.

Bhatti, M. I., and Bodla, M. A. (2013), "On the Mismeasured of Nonested Tests for the Error-in-Variables Models", memo Department of Finance, La Trobe University, submitted for the possible publication.

Bodla, M.A., and Bhatti, M.I. (2008), "Empirical Power Comparison of Nonnested Tests for the EVM: Some Monte Carlo Evidence, International Journal of Applied Econometrics and Quantitative Studies Vol. 5-2, 87-102.

Bodla, M.A. and Bhatti, M. I. (2007), Use of Instrumental Variables in Non-nested EV-Models" Journal of Statistical Theory and Application, Vol. 6(3), 254-263.

Bontemps, C., Mizon, G. E. (2008), "Encompassing: Concepts and Implementation", *Oxford Bulletin of Economics and Statistics*, *70*: 721-750.

Box, G.E.P., Cox, D.R. (1964), "An Analysis of Transformations", *Journal of Royal Statistical Society*, Series B (Methodological), Vol.26 No.2, pg.211-252.

Breusch, T.S. (1979), "Conflict among Criteria for Testing Hypotheses: Extension and Comment", *Econometrica*, 47, pg. 203-208.

Clark, T.E., McCracken, M.W. (2009), "Combining forecasts from nested models", *Oxford Bulletin of Economics and Statistics*, pg. 303–329.

Cox D.R. (1961), "Tests of separate families of hypotheses", *Proceedings of the Fourth Berkeley Symposium on Mathematical Statistics and Probability*, 1, pg. 105-123.

Cox D.R. (1962), "Further results on tests of separate families of hypotheses", *Journal of the Royal Statistical Society*, Series B, 24, pg. 406-424.

Dadkhah, K.M., Valbuena (1985), "Non-nested test of New Classical vs Keynesian model: Evidence from European economies", *Applied Econometrics*, 17, pg. 1083-1098.

Das Gupta, S. and Perlman, M.D. (1974), "Power of the non-central F-test: Effect of the additional variates on Hotelling's T^2-test", *J. Amer. Statistic. Associate*, 69, pg. 174.

Dastoor, N.K. (1981), "A note on the interpretation of the Cox procedure for non-nested hypotheses", *Economics Letters*, 8(2), pg. 113–119.

Dastoor, N.K. (1983), "Some aspects of testing non-nested hypotheses", *Journal of Econometrics*, 21(2), pg. 213–228.

Dastoor, N.K., McAleer, M. (1985), "Testing separate models with stochastic regressors", *Economic Modelling* ,2 (4).

Davidson R., MacKinnon J.G. (1981), "Several tests for model specification in the presence of alternative hypotheses", *Econometrica*, 49, pg. 781-794.

Davidson, R., MacKinnon, J.G. (1982), "Some Further Results on Tests for Model Specification in the Presence of Alternative Hypotheses", *Working Papers 491*, Queen's University, Department of Economics.

Davidson R., MacKinnon J.G. (2001), "Improving the reliability of bootstrap tests", *Queen's University Institute for Economic Research Discussion Paper No. 995*, revised.

Davidson, R., MacKinnon, J.G. (2002), "Bootstrap J tests of non-nested linear regression models", *Journal of Econometric*, 109, pg. 167-193.

Delgado, M., Stengos, T. (1994), "Semiparametric specification testing of non-nested econometric models", *Review of Economic Studies, 61*, pg. 291–303.

Dhrymes, P.J., Howrey, E.P., Hymans, S. H., Kmenta, J., Leamer, E.E., Quandt, R.E., Ramsey J.B., Shapiro, H.T. and Zarnowitz, V. (1972), "Criteria for evaluation of econometric models", *Annals of Economic and Social Measurement*, pg. 291-324.

Dubbelman C., Louter A.S., Abrahamse A.P.J. (1978), "On typical Characteristics of Economic Time Series and the Relative Qualities of Five Autocorrelation Tests", *Journal of Economrtrics*, 8, pg. 295-306.

Econometrica, 41, pg. 997-1016.

Durbin, J. (1954), "Errors in Variables", Revue de l'Institut International de Statistique / Review of the International Statistical Institute, 22,(1/3), 23-32.

Epps, T.W., Singleton, K.J. and Pulley, L.B. (1982), "A Test of Separate Families of Distributions Based on the Empirical Moment Generating Function", *Biometrika*, 69(2), pg. 391-399.

Ericsson, N.R. (1986), "Post-simulation Analysis of Monte Carlo Experiments: Interpreting Pesaran's (1974) Study of Non-nested Hypothesis Test Statistics", *Review of Economic Studies*, LIII, pg. 691-707.

Ericsson, N.R. (1983), "Asymptotic Properties of Instrumental Variables Statistics for Testing Non-Nested Hypotheses", *The Review of Economic Studies*, 50(2), pg. 287-304.

Fan, Y., Li, Q. (1995), "Bootstrapping J-type tests for non-nested regression models", *Economic Letters*, 48, pg. 107-112.

Fernandez, A. Z., Koenig, E. F., and Nikolsko-Rzhevskyy, A. (2010), "Can alternative Taylor-rule specifications describe Federal Reserve policy decisions?", *Journal of Policy Modeling*, 32, pg. 733–757.

Fisher, G.R., McAleer, M. (1979), "On the interpretation of the Cox test in econometrics", *Economics Letters*, 4, pg. 145–150.

Friedman, M. (1957), "Theory of the Consumption Functhon Princeton", N.J.: Princeton Univ. Press.

Fuhrer, J., Moore, G. (1995), "Inflation persistence", *Quarterly Journal of Economics*, 110, pg. 127–160.

Gaver, K.M., Geisel, M.S. (1974), "Discrimination among alternative models: Bayesian and non-Bayesian methods", *In Frontiers of Econometrics (P. Zarembka, ed.)*, New York: Academic Press.

Giacomini, R., Rossi, B. (2007), "Model Selection and Forecast Comparison in Unstable Environments", *(University of California, Los Angeles, and Duke University, January, unpublished paper)*.

Godfrey, L.G. (1998), "Tests of non-nested regression models: some results on small sample behaviour and the bootstrap", *Journal of Econometrics*, 84, pg. 59-74.

Godfrey, L.G. Pesaran, M.H. (1983), "Tests of Non-nested Regression Models: Small Sample Adjustments and Monte Carlo Evidence", *Journal of Econometrics*, 21, pg. 133-154.

Godfrey, L.G., Wickens, M.R. (1981), "Testing Linear and Log-Linear Regressions for Functional Form", *The Review of Economic Studies*, 48(3), pg. 487-496.

Granger, C.W.J. (1966), "The Typical Spectral Shape of an Economic Variable", *Econometrica*, 34, pg. 150-161.

Green, R.D., Pope, R.D., and Phipps, T.T. (1981), "Discriminating among Alternative Habit Formation Schemes in Single-Equation Demand Models", *Applied Economics*, 13, pg. 399-409.

Gregory, A.W., McAleer, M. (1983), "Testing Non-Nested Specifications of Money Demand for Canada", *The Canadian Journal of Economics*, 16(4), pg. 593-602.

Griliches, A. (1974), "Errors in Variables and Other Unobservables", *Econometrica*, 42, pg. 971-998.

Harvey, A.C. (1977), "A Comparison of Preliminary Estimators for Robust Regression", *Journal of the American Statistical Association*, 72, pg. 910-913.

Harvey, A.C. (1981), "Time Series Models", *New York: John Wiley.*

Haslag, G.H., Hein, S.E. (1990), "Economic activity and two monetary base measures", *The Review of Economics and* Statistics, 72, pg. 672-677.

Hoel, P.G. (1947), "On the Choice of Forecasting Formulas", *Journal of the American Statistical Association*, 42(240), pg. 605-611.

Hong, H., Preston, B. (2011), "Bayesian averaging, prediction and non-nested model selection", *Journal of Econometrics*, doi: 10.1016/ j.ecohom.2011.09.021.

Huth, P., Gelpi, C., Bennett, D.S. (1993), "The escalation of great power militarized disputes: Testing rational deterrence theory and structural realism", *American Political Science Review*, 87, pg. 609-623.

Jackson, A.Y. (1968), "Some results on tests of separate families of hypotheses", *Biometrika*, 55, pg. 355-363.

Johannes, J.M., Nasseh, A.R. (1985), "Income or Wealth in Money Demand: An Application of Non-nested Hypothesis Tests", *Southern Economic Journal*, 51, pg. 1099-1106.

Kelejian, H.H. (2008), "A spatial J-test for model specification against a single or a set of non-nested alternatives", *Letters in Spatial and Resources Sciences,* 1(1), pg. 3–11.

Kelejian, H. H., Piras, G. (2011), "An extension of Kelejian's J-test for non-nested spatial models", *Regional Science and Urban Economics*, 41, pg. 281–292.

Khan, H., Zhu, Z. (2002), "Estimates of the sticky-information Phillips curve for the United States, Canada, and the United Kingdom", *Bank of Canada Working Paper 2002*, No. 2002-19.

Kiley, M. (2007), "A quantitative comparison of sticky-price and sticky-information models of price setting", *Journal of Money, Credit and Banking*, 39, pg. 101–125.

King, G. (1989), "Unifying political methodology", Cambridge, *Cambridge University Press.*

King, M.L. (1998), "The likelihood ratio test of non-nested linear regression models".

King, M.L., McAleer, M. (1984), "Further Results on Testing AR(1) Versus MA(1) Disturbances in the Linear Regression Model", *Economics and Econometrics Paper*, Australian National University.

Koenker, R., Bassett G. (1978), "Regression quantile", *Econometrica*, 46, pg. 33-50.

Korenok, O. (2008), "Empirical comparison of sticky price and sticky information models", *Journal of Macroeconomics,* 30, pg. 906–927.

Kuan, C., Lin, H. (2010), "An encompassing test for non-nested quantile regression models", *Economics Letters*, 107, pg. 257–260.

Kullback, S. (1968), "Information Theory and Statistics" (Dover, NewYork).

Kullback, S., Leibler, R.A. (1951), "On Information and Sufficiency", *Annals of Mathematical Statistics*, 22, pg. 79-86.

Lodewyckx, T., Kim, W., Lee, M. D., Tuerlinckx, F., Kuppens, P., and Wagenmakers, E. (2011), "A tutorial on Bayes factor estimation with the product space method", *Journal of Mathematical Psychology*, 55, pg. 331–347.

Mankiw, N.G., Reis, R. (2001), "Sticky information: A model of monetary non-neutrality and structural slumps", *NBER Working Paper 2001*; No. 8614.

McAleer, M. (1995), "The significance of testing empirical non-nested models", *Journal of Econometrics*, 67, pg. 149-171.

McAleer, M., Desitler, M. (1986), "Some Recent Development in Econometrics. Working Papers in Economics and Econometrics", *The Australian National University*, Faculty of Economics and Research School of Social Sciences.

McAleer, M. (1987), "Specification Tests for Separate Models: A Survey", in King, M.L. and D.E.A.Giles (Eds) *Specification Analysis in the Linear Model*, pg. 146-195.

Mizon, G. E. (1984), "The encompassing approach in econometrics", in Hendry D. F. and Wallis K. F.(eds), *Econometrics and Quantitative Economics*, Basil Blackwell: Oxford, pg. 135–172.

Mizon, G.E., Richard, J.F. (1986), "The Encompassing Principle and its Application to Testing Non-Nested Hypotheses", *Econometrica*, 54(3), pg. 657-678.

Morimune, K. (1979), "Comparisons of Normal and Logistic Models in the Bivariate Dichotomous Analysis", *Econometrica*, 47(4), pg. 957-975.

Morimune, K. (1983), "Approximate Distributions of k-Class Estimators when the Degree of Overidentifiability is Large Compared with the Sample Size", *Econometrica,* 51(3), pg. 821-84.

Otsu, T., Seo, M.H., and Whang, Y. (2011), "Testing for non-nested conditional moment restrictions using unconditional empirical likelihood", *Journal of Econometrics*, doi: 10.1016/j.econom.2011.09.022.

Pagan, A.R., Hall, A.D., Trivedi, P.K., "Assessing the Variability of Inflation", *Reviews of Economic Studies*, 50(1983), pg. 585-596

Pereira, B. DE B. (1977), “A Note on the Consistency and on the Finite Sample Comparisons of Some Tests of Separate Families of Hypotheses”, *Biometrika*, 64(1), pg. 109-113.

Pereira, B. DE B. (1978), “Empirical Comparisons of Some Tests of Separate Families of Hypotheses”, *Metrika*, 25, pg. 219-234.

Pesaran, M.H., Pesaran, B. (1993), “A simulation approach to the problem of computing Cox’s statistic for testing non-nested models”, *Journal of Econometrics*, 57, pg. 377-392.

Pesaran, M.H. (1982), “Comparison of local power of alternative tests of non-nested regression models”, *Econometrica,* 50, pg. 1287–1305.

Pesaran, M.H. (1982a), “On the comprehensive method of testing non-nested regression models”, *Journal of Econometrics*, 18, pg. 263–274.

Pesaran, M.H. (1987), “The Limits to Rational Expectations”, Oxford, Blackwell.

Pesaran, M.H. (1984), “Asymptotic power comparisons of tests of separate parametric families by Bahadur's approach”, *Biometrika*, 71, pg. 245-252.

Pesaran, M.H., Deaton, A.S. (1978), “Testing Non-Nested Nonlinear Regression Models”, *Econometrica*, 46, pg. 677-694.

Pesaran, MH. (1987), “The Limits to Rational Expectations”, Oxford: Basil Blackwell.

Pesaran M.H. (1974), “On the General Problem of Model Selection”, *The Review of Economic Studies*, 41(2), pg. 153-171.

Piras, G., Lozano-Gracia, N. (2010), “Spatial J-test: some Monte Carlo evidence”, *Statistics and Computing*, pg. 1–15.

Quandt, R.E. (1974), “A Comparison of Methods for Testing Non-nested Hypotheses”, *The Review of Economics and Statistics*, 56(1), pg. 92-99.

Ramsey, J.B. (1974), “Limiting Forms for Demand Functions: Tests of Some Specific Hypotheses”, *The Review of Economics and Statistics*, 56(4), pg. 468-477.

Sbordone, A. (2002), “Prices and unit labor costs: A new test of price stickiness”, *Journal of Monetary Economics, 49* (2), pg. 265–292.

Shen, S.M. (1982), “A method for discriminating between models describing compositional data”, *Biometrika*, 69, pg. 587-595.

Shumway, C.R., Gottret, P.E. (1991), “Numeraire choice in normalized quadratic functional forms”, *Applied Econometrics*, 23, pg. 1597-1602.

Silvapulle, P., King, M.L. (1993), “Non-nested testing for autocorrelation in the linear regression model”, *Journal of Econometrics*, 58, pg. 295-314.

Smith, M.A., Maddala, G.S. (1983), "Multiple model testing for non-nested heteroskedatic censored regression models", *Journal of Econometrics*, 21, pg. 71-82.

Thornton, D.L. (1985), "Money Demand Dynamics: Some New Evidence", *Federal Rexme Bank of St. Lob Reuiew 67*, (March) 14-23.

Wald, A. (1940), "The fitting of straight lines if both variables are subject to error", Annals of Mathematical Statistics 11, 284–300.

White, H. (1982), "Regularity conditions for Cox's test of non-nested hypothesis", *Journal of Econometrics*, 19, pg. 301–318.

Wisely, T.O., and Johnson, S.R. (1985), "An Evaluation of Alternative Investment Hypotheses Using Non-Nested Tests", *Southern Journal of Economics*, 52, pg. 422-430.

Wright, J.H. (2008), "Bayesian model averaging and exchange rate forecasts", *Journal of Econometrics*, 146 (2), pg. 329–341.

Zellner, A. (1971), "An Introduction to Bayesian Inference in Econometrics", New York: Wiley.

Zellner, A. (1984), "Basic issues in econometrics", *University of Chicago Press*, Chicago, IL. pg. 187-200.

INDEX